INVERTEBRATE LEARNING

Volume 2
Arthropods and Gastropod Mollusks

INVERTEBRATE LEARNING

Volume 2
Arthropods and Gastropod Mollusks

Edited by

W. C. Corning and J. A. Dyal

Department of Psychology
University of Waterloo
Waterloo, Ontario, Canada

and

A. O. D. Willows

Department of Zoology
University of Washington
Seattle, Washington

PLENUM PRESS · NEW YORK-LONDON · 1973

Library of Congress Catalog Card Number 72-90335

ISBN 0-306-37672-5

© 1973 Plenum Press, New York
A Division of Plenum Publishing Corporation
227 West 17th Street, New York, N. Y. 10011

United Kingdom edition published by Plenum Press, London
A Division of Plenum Publishing Company, Ltd.
Davis House (4th Floor), 8 Scrubs Lane, Harlesden, London, NW10 6 SE, England

CONTENTS OF VOLUME 2

Chapter 9
Honey Bees
Patrick H. Wells

Chapter 10
Learning in Gastropod Mollusks
A. O. D. Willows

CONTENTS OF VOLUME 1

Chapter 3
Behavioral Modifications in Coelenterates
N. D. Rushforth

$Vol. I.$

Chapter 4
Platyhelminthes: The Turbellarians
W. C. Corning and S. Kelly

Chapter 5
Behavior Modification in Annelids   Vol. I.
  J. A. Dyal

CONTENTS OF VOLUME 3

Chapter 6

THE CHELICERATES

Robert Lahue

Department of Psychology
University of Waterloo
Waterloo, Ontario, Canada

I. INTRODUCTION

Of approximately 900,000 arthropod species, the subphylum Chelicerata accounts for about 60,000. The absence of jaws and antennae has served as the primary point of distinction between the chelicerates and the remaining arthropod subphylum, the Mandibulata. The invasion of all possible terrestrial habitats has been aided by the chitinous exoskeleton, which protects against desiccation, and by the jointed appendages, which permit rapid locomotion on land. The phylum as a whole, and many species in particular, represents the peak of protostome evolution, especially of the nervous system.

The Chelicerata (Table I) provide an exciting array of specimens for both comparative neurophysiologists and comparative psychologists. The aquatic *Limulus* (class Merostomata) is quite similar to its ancestors of 200 million years ago and has changed little if at all in over 75 million years. As such, *Limulus* provides a living link between all terrestrial arthropods and the past. Of all recent terrestrial animals, the Arachnida are probably the oldest class, with scorpions occupying an intermediate position between *Limulus* and the highly advanced spiders.

There is general agreement concerning the origin of the Chelicerata with the trilobites. The class first appeared in the late Pre-Cambrian. At about the same time as the initial stages of chelicerate evolution, the Upper Cambrian representatives of the Merostomata appeared. In fact, the first Chelicerata

1

Table I. Classification[a]

Phylum Arthropoda	
Subphylum Chelicerata	Arthropods lacking mandibles and antennae
Class Merostomata	Primitive aquatic chelicerates
Order Xiphosura	*Limulus polyphemus, Tachypleus gigas, Carcinoscorpius*
Order Eurypterida	Extinct sea scorpions
Class Arachnida	Terrestrial chelicerates
Order Scorpionida	Scorpions
Family Buthidae	*Androctonus, Buthus, Buthotus*
Family Scorpionidae	*Scorpio*
Order Araneae	Spiders
Family Salticidae	Jumping spiders—good vision, little use of silk: *Salticus scenicus, Evarcha blancardi, Epiblemum scenicum*
Family Lycosidae	Wolf spiders—large wandering spiders, fair vision: *Arctosa variana, A. cinerea, A. perita, Lycosa fluviatiers, Aphonopelma californica, Phidippus morsitans, Astia vittata, Xysticus ferox*
Family Pisauridae	Wolf-spiders: *Dolomedes fimbriatus*
Family Agelenidae	*Agelena labyrinthica, Tegenaria domestica*
Family Theridiidae	*Achaearanea tepididariorum*
Family Argiopidae	Highly evolved, aerial web spinners: *Araneus diadematus, Zygiella x-notata*
Family Linyphiidae	*Linyphia triangularis*

[a]Examples provided are of species discussed in the text. Not all orders are included.

probably also belonged to the Merostomata (Raw, 1957; Sharov, 1966). The order Xiphosura are the only extant members of the class Merostomata.

The traditional view holds that the Scorpionida, appearing in the late Ordovician, arose from the Eurypterida (sea scorpions), which branched from the Merostomata in the Upper Cambrian. However, Störmer (1963) claims that the Scorpionida and Eurypterida are linked only by a common ancestor, and the problem has not yet been resolved.

Also arising from the Eurypterida, probably earlier in the Ordovician than the Scorpionida, were the Pedipalpida. The evolution of pedipalpids from eurypterids apparently repeated in parallel fashion scorpion evolution. Three branches derived from the Pedipalpida: the Araneae, Solifugida, and Acarina. The Araneae are probably the class most similar to the pedipalpids.

II. GENERAL CHARACTERISTICS

A. Habitat

The three existing genera of the Merostomata live in shallow coastal waters and are primarily aquatic, coming on land only to lay eggs. *Limulus* inhabits the waters of the East Coast of North America from Nova Scotia to Mexico. *Tachypleus* is widely distributed from Malaysia to southern Japan, while *Carcinoscorpius* inhabits the shores of the Indian Ocean.

Arachnida are terrestrial arthropods found on all the continents. Although the class as a whole is widely dispersed, many orders such as the scorpions are confined to tropical and subtropical regions. While theraphosid and liphistiid spiders are also restricted to such warm areas, the majority of spiders are found throughout the temperate zones. Araneae are found at 22,000 ft on Mount Everest and at the northernmost tip of Greenland. Mites have been found in the Antarctic.

The majority of the Arachnida live a concealed mode of life and rarely travel far. They choose to live in microenvironments which produce the greatest degree of relative constancy: crevices, underground burrows, or under stones and layers of leaves. The physiological and structural characteristics of these animals, which are representative of the primitive adaptations to terrestrial life made by their aquatic ancestors, are the basis for such restrictions. Relatively minor environmental variations are greatly magnified with respect to the physiology of arachnids. The few species able to survive in a less constant environment have evolved special adaptations. Thus many web-building spiders are found to possess a thicker cuticle than do other arachnids.

B. General Morphology

The Chelicerata may be distinguished from the other arthropods by the absence of jaws and antennae. The body is divided into two parts. The anterior part, the prosoma, consists of a fused head and thorax and is often referred to as the cephalothorax. The remainder of the body, the opisthosoma, is commonly referred to as the abdomen. The cephalothorax consists of the first eight embryonic segments, with the first and eighth being absent in the adults of most classes. Each remaining segment carries a pair of appendages, although they are fused dorsally to form a carapace. The abdomen is composed of from six to twelve segments, although spiders retain only four in the adult form.

In *Limulus*, the fusion of the eight cephalothoracic segments has resulted in a semicircular carapace with steeply sloping sides enclosing the appendages. This characteristic shape has yielded the common name of "horseshoe crab" for this species. Two simple eyes (ocelli) are borne anteromedially on the dorsal surface, while the well-known compound eyes are located laterally. Other primitive or degenerative photoreceptive structures are also located dorsally. Several dorsal spines may be noted, and delicate hairs fringe the posterolateral margin. The appendages, which cannot be seen unless the animal is viewed ventrally (see Fig. 1A), are equipped with pincers and are chelate.

The opisthosoma is a broad, flat fusion of six somites with a noticeable trace of segmentation. Six small, movable spines are articulated between seven fixed spines on the posterolateral margins. The long, spined telson is articulated posteromedially. The six pairs of appendages are semicircular, flattened plates. The first and largest pair is the genital operculum, which lies over the simple genital orifices, as well as overlapping one or two pairs of posterior appendages. The next five pairs of appendages regularly overlap one another but are individually movable. Each appendage carries a respiratory organ, or gill book, dorsally on its forward edge. Each of the ten gill books contains between 100 and 200 oval leaves in which the blood flows.

Scorpions are rather primitive arachnids, probably good examples of the early adaptations made to terrestrial life by their eurypterid sea scorpion ancestors. In many characteristics, they present a nice intermediate form between the very old *Limulus* and the relatively recent and highly adapted arachnids such as spiders (see Fig. 1B).

Certain features serve to identify scorpions, regardless of size, and distinguish them from other arachnids. The pedipalps are quite large and furnished with very strong claws (chelae). The abdomen is divided into two easily discernible sections: an anterior mesosoma or preabdomen, and an elongated metasoma or postabdomen. The preabdomen consists of seven segments and is usually wider than the cephalothorax, while the postabdomen is slender and tail-like and terminates in a curved, pointed sting. Four pairs of book lungs are located posteriorly to the pectines on the ventral surface.

In contrast to the abdomen, the cephalothorax shows little evidence of dorsal segmentation. One pair of median eyes and three to five pairs of lateral eyes are all ocelli.

In Araneae, the eight cephalothoracic segments have fused into a uniform carapace with little if any trace of segmentation (see Fig. 1C). The ocular region is elevated in some species, and some or all of the simple eyes are located there. Most species have eight eyes, although some species have only two eyes, while others have none. In many species, the eyes are of more than one type and are capable of specialization.

The opisthosoma of spiders is usually cylindrical or egg-shaped and shows no signs of segmentation. Sometimes, elaborate patterns are distin-

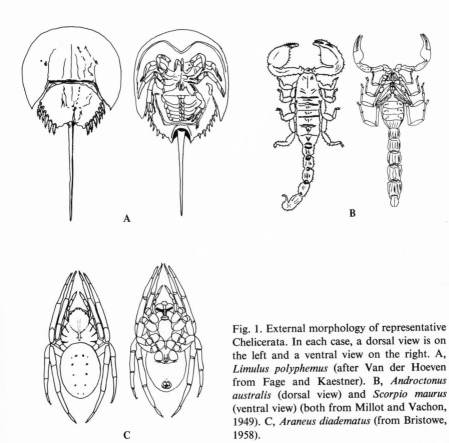

Fig. 1. External morphology of representative Chelicerata. In each case, a dorsal view is on the left and a ventral view on the right. A, *Limulus polyphemus* (after Van der Hoeven from Fage and Kaestner). B, *Androctonus australis* (dorsal view) and *Scorpio maurus* (ventral view) (both from Millot and Vachon, 1949). C, *Araneus diadematus* (from Bristowe, 1958).

guishable on the surface. In some families, the abdomen is modified into extreme forms, the value or purpose of which is often inconceivable. The genital orifice and respiratory openings lie on the ventral surface. The only abdominal appendages remaining in adult spiders arise from the fourth and fifth opisthosomatic segments and function as spinning organs. The number of spinnerets and their size and structure vary among families and are related to the type of web built.

C. Nervous System—General

The nervous system of chelicerates is composed of a number of paired ganglia, one pair in each embryonic segment, which migrate anteriorly and fuse to a greater or lesser extent in the adult animal. All orders possess a typical protocerebral mass which lies dorsally with respect to the pharynx and is usually referred to as a supraesophageal ganglion. This mass receives input from the eyes and frontal organs. Several discrete structures are usually discernible.

The anterior optic lobes are composed of several masses of neuropile, each surrounded by small globuli cells. Axons from the eyes seem in all cases to make synapses upon entering the brain. Secondary fibers project ipsilaterally to other adjacent neuropiles as well as to the central body and corpora pedunculata.

The central body is a mass of neuropile situated medially. The only available correlation between size and function is in the Araneae, in which increasing complexity of web structure, and presumably tactile sense, is associated with extremely large central body mass.

Mushroom bodies (corpora pedunculata) are also present in most species, although virtually absent in some web-spinning spiders. As mentioned above, these areas receive input from higher-order visual centers. Although some spider species with complex behavior patterns dependent on memory of visual cues have relatively well-developed corpora pedunculata, correlation between size of these bodies and complexity of behavior has not been adequately demonstrated. For example, the extreme size of these bodies in *Limulus* has yet to be explained.

A deutocerebrum, as such, does not exist in any chelicerates, because of their characteristic lack of antennae. The tritocerebrum lies ventral to the

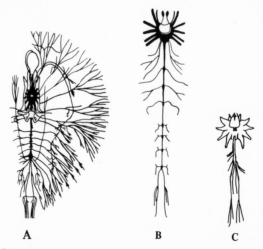

Fig. 2. Structure of the central nervous system of representative Chelicerata. A, *Limulus polyphemus* (after Patten and Redenbaugh, 1900). B, Scorpions in general (from Millot, 1949). C, *Tegenaria domestica* (from Kaestner, 1968).

forebrain and alongside and behind the mouth. In *Limulus,* the ganglionic masses of segments 2–8 migrate forward and fuse with the tritocerebrum to form a subesophageal ganglion innervating abdominal segments 7 and 8 as well as the cephalothorax (see Fig. 2A). The remaining ganglia, innervating segments 9–14 or 9–15, are free in the opisthosoma, lying at intervals along a ventral nerve cord through which they communicate with each other and with the brain. Scorpions also possess a long ventral cord with seven free abdominal ganglia (see Fig. 2B). All ganglia have migrated and fused anteriorly in most spiders (see Fig. 2C). the exception being Mygalomorphae, which possess one free terminal ganglion.

One additional feature of the chelicerate nervous system which distinguishes it from that of all other arthropods is that it is entirely surrounded by an arterial sheath which allows it to be constantly perfused by circulating blood.

D. Nutrition

Limulus is the only chelicerate which eats solid food. The diet consists mainly of worms, small soft-shelled clams, shrimp, and crabs. Movement of the legs causes the food to be crushed by the gnathobases which surround the mouth. When the food has been converted to a pulpy form, the chelicerae push it into the mouth. All other Chelicerata ingest food in a liquid form. This form of nutrition is necessitated by the characteristic absence of jaws. The main food of spiders and scorpions consists of a wide variety of other arthropod forms. Poison inflicted with a bite from the fangs usually immobilizes or kills the prey. The gnathobases of the scorpion pedipalps chew the victim (such structures are nonexistent in spiders), while enzymes secreted from the salivary glands predigest the food before ingestion. The liquid food then rises into the esophagus partly by surface tension and partly by a sucking action of the stomach.

E. Mating and Reproduction

In all Chelicerata, the sexes are separate and, with the exception of *Limulus,* the ova are internally fertilized. This is usually accomplished at the time the eggs are laid, the spermatozoa having been stored within the female previously. The process which results in the transference of the spermatozoa from the male to the female varies considerably among the different orders and in some cases is quite elaborate.

As an aquatic creature, *Limulus* does not require complex fertilization techniques. At high tide, the female, with the male attached, proceeds to the water's edge where she digs a shallow hole in the sand into which 200–300 eggs are deposited, immediately fertilized by the male, and buried. There is no need for courtship, since the male seems to be able to identify the female by some chemical secretion. Frequently, the process is repeated in several locations. Before the tide ebbs, the pair leaves the nest never to return.

When the terrestrial mode of life was assumed, the problem of reproduction became immensely more difficult, and the arachnids developed several means for safely achieving fertilization. The morphological and behavioral adaptations vary widely among the orders and are probably in their most advanced form in the Araneae.

Scorpions have developed a system of fertilization which can probably be judged as intermediate in complexity between that of *Limulus* and that of spiders. A spermatophore is formed in special organs (paraxial organs) and filled with sperm. When its formation is complete, it is cemented to a suitable substrate. The spermatophore consists of a stalk, an ejection apparatus, and a trigger lever. The female presses down on the lever and causes the sperm to be ejected into the gonopore. Both the structure of the paraxial organs and the details of the spermatophore are of systematic value.

In the spiders, the adult male pedipalpal tarsus is specially modified as an intromittent organ as well as a reservoir for temporary sperm storage. The spermatozoa are transferred, via differing mechanisms, from the genital orifice to these organs, from which they may be deposited in the female. The aperture for the female oviduct is larger than the male orifice and is closely associated with one or two receptacles which receive and store sperm. These and other structures which function as ovipositors are covered by an operculum.

Although in some arachnids courtship is nonexistent or perfunctory at most, this is certainly not the usual state of affairs. It would seem that courtship is a necessary means of delaying insemination until intromission may be precisely and efficiently executed.

The essentials of courtship for scorpions are two. Having met and engaged a female, the male searches for a suitably firm substrate for deposition of the spermatophore, resulting in the usual courtship dance. The pectines are probably used to detect substrate suitability. When the spermatophore is ready, the female is guided over it until she senses it, again with the pectines. The genital operculum then opens and the female settles upon the spermatophore, triggering it as a consequence.

In spiders, there are a wide variety of courtship behaviors, the nature of which is determined for any species by the physical characteristics of that species. The primitive wandering spiders manage a simple copulation with a minimum of preliminary stroking. In *Salticidae* and *Lycosidae,* spiders possessing particularly good eyesight, the female may recognize the male by species-specific patterns of coloration of the legs, which are waved before her semaphore-like in fixed sequences of motion. Web spiders, less noted for visual acuity, are extremely sensitive to vibrations reaching their legs through the web. The male of the species wanders upon the web of a female and by characteristic patterns of vibration distinguishes himself from prey, after which he cautiously approaches and eventually mates. Contrary to popular opinion, it is probably rather rare for the female to devour the male after mating, although this is often the case in certain species, *Araneus diadematus,* for example.

F. Respiration

As previously described, oxygenation is carried out in *Limulus* by the circulation of the blood through the gills, which are exposed to the water except for the movable gill book covers. In the Scorpionida and Araneae, the gills have become modified and internalized. Invaginations of the exoskeleton result in a number of hollow folds of delicate epithelium which project into the lung cavity. Air reaches the blood through these epithelial leaves after entering the body through pulmonary spiracles. There may be one, two, or four pairs of book lungs with from five to 150 leaves in each.

Tracheae, respiratory devices quite common in other arthropods, are also present in many spiders. They are narrow chitinous tubes which are extensively branched. Air enters by simple diffusion at stigmata in various locations. With some exceptions, spiders possess both kinds of organs in varying numbers, while scorpions do not possess tracheae.

The blood is an almost colorless fluid. In some spiders, the blood itself is known to be toxic to small mammals. Lying in a pericardium dorsomedially in the opisthosoma is a tubular heart with arteries leaving it in both directions. After passing through smaller vessels, the blood reaches the gills or lung books. It collects in a ventral sinus after aeration and reenters the pericardium through valved ostia. The number of pericardial ostia may vary, and the number among Araneae is greater than in any other order.

G. Sense Organs

Chelicerate eyes are usually of the simple type (ocelli). The only exception is the compound lateral eye of *Limulus,* although the eurypterid scorpions apparently possessed similar structures. These compound eyes have provided a popular preparation for physiologists; their structure and function have been reviewed elsewhere (Wolbarscht and Yeandle, 1967). When correlated with the relatively complex structure and function of the optic areas of the brain (Snodderly, 1969, 1971), the capabilities of these eyes are quite remarkable considering the rather limited use the animal makes of visual information. In fact, very little is actually known about what *Limulus* sees, although techniques for chronic monitoring of optic nerve activity have been developed (Corning *et al.,* 1965). Adolph (1971) has recently recorded optic nerve activity of freely moving crabs under water.

Lying dorsally near midline are another pair of light receptors, the median eyes, which are simple in structure. Functionally, they must contribute little more than simple photoreception.

The eyes of scorpions have been studied little. Both the median and lateral eyes are of the simple type, although they differ structurally. Abushama (1964) has indicated that they do function in predatory behavior.

On the other hand, the eyes of the Araneae have been relatively well studied (Homann, 1928, 1931, 1934; Kaestner, 1950; Dzimirski, 1959; Young and Wanless, 1967; Land, 1969*a,b*, 1971). They are ocelli in all cases, with a wide variety of structural as well as functional features. Typically, the anterior pair of median eyes is most clearly differentiated structurally and functionally from the others, which may number four pairs. In some cases, the angle of sight of the eye or the focal length of the retina may be altered by muscular contractions (Dzimirski, 1959; Land, 1969*a*).

Several other types of sense organs are characteristic of the Chelicerata, but, in general, even less is known about these. *Limulus* is covered in many ventral areas with sensory hairs which are quite sensitive to light pressure. The same or similar hairs line the pitlike area in which the lateral movable spines are articulated and function as indicators of movement in the spine. There is also a neural mechanism (the details of which are unknown at this time) which detects displacement of the spine in the absence of the hairs being stimulated. Mechanoreceptors are present in the leg joints and are especially accessible for recording in the coxa–trochanter joint. Single-neuron sensilla located on the gnathobasal spines and chelae are probably also mechanoreceptors. Chemoreceptors are located on the mouth and parts of the surrounding appendages. The gnathobasal spines respond to chemical, thermal, and tactile stimuli.

Very little is known about the sense organs of scorpions. As in other arachnids, numerous sensory hairs (setae), trichobothria, and slit sense organs are present. Trichobothria are rather complex organs consisting of long, delicate, movable hairs which are located in a joint membrane. In scorpions, the trichobothrial setae can move in only one plane, although the plane of movement of different hairs varies. In spiders, movement of each hair is in many planes and the innervation is more complex. It has been suggested that these function as auditory organs, since they respond to air vibrations. However, this has not been proven, and in scorpions at any rate they are probably detectors of slight wind currents. It is also possible that they function at least in some cases as chemo- and thermoreceptors. These trichobothria occur on the pedipalps and all the legs of spiders but only on the former in scorpions.

Slit sense organs (lyriform organs) located on the legs, carapace, and sternum appear as slits in the cuticle and occur singly or in groups. Although they seem to function as vibration receptors, chemoreceptive and mechanoreceptive functions are probably also present; however, this has yet to be demonstrated adequately. In *Achaearanea,* individual organs on the legs are tuned to certain ranges of frequency, and each leg through several receptors is sensitive in the range 20–45,000 cps (Walcott and Van der Kloot, 1959). Tuning of each organ can be modified by leg flexion, and web spiders orienting to vibrating stimuli are often seen altering leg flexion as if to tune in the particular vibration frequency present. The function of these organs in scorpions has not been reported.

Scorpions possess comblike remnants of the anterior abdominal appendages called pectines which are typified by complex patterns of sensory structures on their surface. They probably function as mechanoreceptors sensitive to vibratory stimuli as well as touch. Evidence is available which indicates that they also detect substrate texture and humidity.

Other than the tarsal organs of spiders, little is known concerning chemical senses in arachnids, although experiments clearly indicate that certain chemicals can influence behavior.

III. LEARNING STUDIES

A. Habituation

Numerous early investigators reported instances in which the behavioral response to some stimulus gradually waned with repeated presentation of the stimulus. Since all of the reports are anecdotal in nature and rarely included any controls, only one of the more interesting older studies will be outlined here.

A struck tuning fork held near an orb-weaving spider will cause it to drop from its web on a thread. The Peckhams (1887) found that with repeated presentations of the tuning fork the distance which the spider dropped decreased until no response was elicited. On succeeding days, the maximum response and the number of responses elicited were lower than on the initial day of training. Although there was great day to day and subject variability, as long as training was continued the response decrement was retained. Wen (1928) reported results quite similar to the Peckhams' regarding total number of acquisition trials and amount of retention. Savory (1934) noted that the habituated response spontaneously recovered to a new stimulus (e.g., a tuning fork of a different frequency), although some generalization was observed. Other authors reported similar findings with a variety of stimuli and responses (Barrows, 1915; Murphy, 1914, Thomas, 1913; Berland, 1934).

Recently, the defensive withdrawal response in some scorpions has been shown to habituate to repeated stimulation, although the closely related strike response is quite resistant to the same process (Palka and Babu, 1967).

1. Predatory Behavior in Jumping Spiders

Spiders of the family Salticidae provide excellent specimens for behavioral studies. All salticids are diurnal spiders possessing excellent vision. Prey are actively hunted, and during courtship the male displays elaborately.

These "jumping spiders" are the most advanced of the vagrant spiders, which make little use of silk. They are certainly the best sighted of spiders and probably one of the most visually dependent of all animals (Land, 1969a). One often notices a salticid turning the cephalothorax to view an object while the abdomen remains fixed in position. Frequently, one of these rather small creatures (few species attain more than several millimeters in size) will sit on the edge of a table or even on a hand, turning the cephalothorax to and fro to follow one's every movement.

Characteristic of all species is the arrangement of the eyes in three rows. The four largest eyes are in a single row across the front of the cephalothorax, with two pairs of smaller eyes located somewhat more posteriorly. The anterior eyes are capable of receiving a sharp image at a distance of several inches and certainly perceive form at such distances. At longer distances, movement is detected by all eyes. The spider always orients toward objects with the anterior eyes.

Land (1969b) described several eye movements in salticids, one of which is of particular interest here. The presentation of a small target (e.g., 3° black dot) with an ophthalmoscopic device upon an outlying region of an anteromedial or anterolateral eye elicits a rapid saccadic movement which brings the image to a central retinal region. Repeated presentation of the target upon the same site at 20-sec intervals results in a decrease in saccadic movements followed by a complete failure to respond. However, a spontaneous recovery of such movements follows stimulation of a different retinal area. In a more recent study, Land (1971) repeated this demonstration in unrestrained animals. The habituated response remained low for periods of several hours. But again movement of the stimulus to a new site resulted in response recovery, as did interpolation of a mechanical stimulation of the body.

If moving objects are detected behind the spider by the posterior eyes, the cephalothorax is turned to bring the objects into the visual field of the anterior eyes. Once an object is viewed by the anterior eyes, the rest of the body turns and orients toward it. The spider then runs up to within 5 cm of the prey. This run is followed by a slow, creeping form of locomotion which brings the spider within jumping range and is in turn followed by a jump. When the object is a female, the slow creeping is replaced by a waving of the front legs and a courtship dance replaces the jump (Drees, 1952; Precht, 1952; Precht and Freytag, 1958; Gardner, 1965).

Since the behavior of these spiders is so largely dependent on visual information, much of the apparatus used in the study of orientation provides useful means of working with behavior. Figure 3 shows several types of equipment which have been used. A simple device described by Heil (1935) is shown in Fig. 3A. The spider is placed in the box, and patterns attached to *A* may be given vertical movement. The device shown in Fig. 3B is again of Heil's design. Although basically the same as A, this apparatus allows the box to be displaced horizontally, thus requiring the spider to jump to the target. A dead fly (*D*) can be moved with the wire *F*.

Drees (1952) employed the device shown in Fig. 3C, which simulated prey movement in a horizontal direction. The device shown in Fig. 3E, also designed by Drees, provided a means of applying electrical shock to a spider which had landed on the target. Gardner (1964) described a clever means of introducing spider and live prey into a closed environment for observation, which is shown in Fig. 3F. A spider held captive in a bottle could be introduced through *A*, which was then slid over to prevent escape. Prey were placed in a shallow depression (*B*) in another sliding floorboard and covered with a glass slip. When the board was slid, the glass was stopped by the side of the box while the prey (flies) could move into the chamber.

In a qualitative sense, Heil (1935) experimentally detailed the determinants of prey orientation in *Evarcha*. Tactile and chemical cues appear to have no role in the orientation, although Bristowe (1958) reported that the behavior of males is influenced by some chemical factor associated with the female.

As previously mentioned, there is a rather well-developed visual sense with several interesting properties. Two-dimensional targets elicit orientation only if moving. However, the dependence of the response on motion is lessened as the target takes on a more complex form. Three-dimensional objects elicit responses rather frequently, and if the object is a dead fly a response always occurs.

A more detailed analysis of these behavioral determinants was undertaken by Drees (1952) using *Epiblemum scenicum*. The general procedure for all of his experiments required that the animals be fed to satiation and then kept in a dark room at 15°C for a variable number of days, after which they were removed to a lighted room kept at 25°C for a couple of hours prior to testing.

Any three-dimensional or two-dimensional target will elicit a response

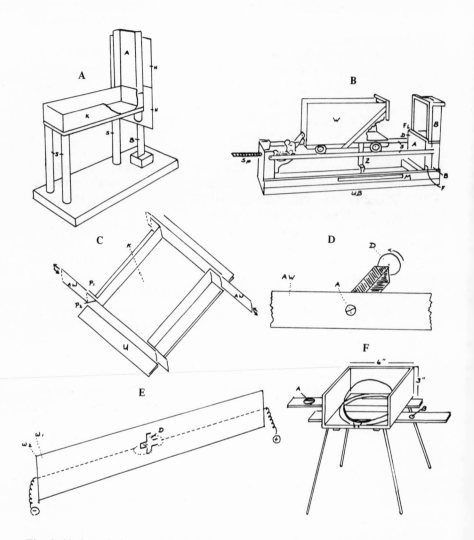

Fig. 3. Various devices used to test the components of prey orientations in salticid spiders. A. While the spider is in the chamber (*K*), targets on *A* may be moved vertically (from Heil, 1935). B. The chamber (*W*) containing the spider may be displaced horizontally from the target (*D*) (from Heil, 1935). C. Movable strips at either end of the test chamber (*K*) can be used to present targets moving horizontally (from Drees, 1952). D. Another device for simulating horizontal movement (from Drees, 1952). E. A movable strip which can be used in (c). When the spider lands on the target (*D*), a shock is received (from Drees, 1952). F. Test chamber described in the text (from Gardner, 1964).

when moving at 2 cm/sec at a distance of 6 cm from the spider. At this fast speed, the creeping portion of the behavioral chain is omitted. Motionless objects and targets are ineffective stimuli at distances greater than 5 cm. Shorter distances facilitate orientation to objects but rarely to targets.

When objects and targets were presented at slow velocities, various behaviors were found to be dependent on speed. A speed of 1 cm/sec produced essentially the same results as faster speeds, eliciting reactions at distances as great as 10 cm. The creeping portion of the advance was still omitted. Halving the velocity only served to decrease the effective distance. Even slower speeds (0.25 cm/sec) reduced the effective distance to 4 cm and elicited the same behavior as motionless objects. Repeated presentations of the stimulus produced interesting results. Spiders were kept under normal pre-experimental conditions for 10 days before being presented with two-dimensional circular black targets moving 2 cm/sec at a distance of 5 cm. The size of the target, which was presented repeatedly, varied between 1, 2, and 3 cm in diameter. Although the rate of presentation was not specified, it certainly must not have been more than once every several seconds, since the animal required time to cease responding to the previous presentation.

Each of eleven subjects was tested in four different situations, as depicted in Fig. 4A. In each situation, target *A* was repeatedly presented until no further jumps were elicited by its movement. Target *B* was then presented until the response to it also disappeared. Although not specified, it is assumed that the situations were presented in a random order to the spiders and that a 10-day rest period intervened between testing in each situation. Although there is a large amount of subject variability both within subjects and across groups, the mean number of jumps elicited by the first target is remarkably consistent. As the preliminary experiments had predicted, the size of the rapidly moving objects did not influence performance. After cessation of the response, however, the size of the *B* target seemed to become an important variable. Thus a small target followed by a larger one elicited more jumps than did the reverse situation.

Further experiments demonstrated the degree to which the animal could respond differentially to such target variables as form, size, and relative complexity. In general, the more a target resembled a natural object (e.g., a fly) the greater was the response elicited by it. As shown in Fig. 4C, the amount of disparity between preferred qualities in two targets influenced the relative response magnitude to each member of the pair when presented simultaneously. Figure 4B demonstrates that abstract targets can be designed to be as effective as more naturally shaped targets.

The experiment summarized in Fig. 4B further elucidates the specificity of the mechanism underlying the response decrement in previous experiments. Figure 4A showed that there was a unidirectional effect of target size

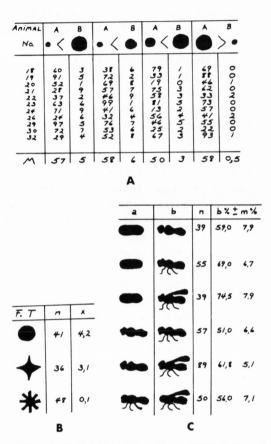

Fig. 4. Habituation of jumping in Salticidae (from Drees, 1952). A, In each situation, the number of jumps elicited by repeated presentation of target *A* is indicated for each spider. The response to subsequent presentation of target *B* is also indicated. B, After habituation to a target (F.T.), presentation of a dummy fly elicits an average of *x* responses. Training on more complex targets (those which resembled a fly) is generalized to the flylike object. C, When targets *a* and *b* are presented simultaneously, the average number of jumps directed to *b* ($b\% \pm m\%$) is dependent on the relative disparity between *a* and *b*.

on the number of responses evoked by the second target. The greatest number of *B* responses resulted when *B* was larger than *A*, and greater differences in size were differentially effective. In the experiment depicted in Fig. 4B, the *A* targets pictured in column 1 are approximately equal in value with respect to size but vary in complexity. The *B* target was a natural-appearing one (as in Fig. 4C). As might be expected, the spider was least able to differentiate between *B* and the third *A* target, the reverse being true of the first *A* target.

The largest B responses followed habituation to the least similar A target. The most similar A target was virtually ineffective. Thus *Salticus* exhibits a remarkable degree of specificity in its response to unrewarding stimuli.

Drees noticed in the course of his experiments that various distinct portions of the prey-catching behavior (i.e., orientation, run, crawl, and jump) were often influenced differentially by his treatments. When he used rapidly moving targets, the crawling segment frequently was not performed. Because of this, Drees used jumping as the criterion for all of his experiments.

However, even though the jumping response had waned, other behavioral responses remained intact. Thus the orientation and initial run were still elicited. With slower-moving targets, the creeping response also persisted. Repeated presentations of a target resulted in the eventual loss of the remaining components, orientation being most resistant to exhaustion.

Precht (1952) and Precht and Freytag (1958) continued with a more detailed analysis of this phenomenon, emphasizing the recovery time and effects of repeated testing. In *Epiblemum scenicum* and *Evarcha blancardi,* some interesting extensions of Drees' procedure were employed which allowed differential elicitation of certain behavioral components. A three-dimensional object lying on the floor of the experimental chamber was made to move in response to a magnet under the floor. Rapidly moving the object at a velocity just equal to the speed of the spider resulted in the spider running after the object but never getting close enough to initiate a jump. A relatively slow-moving or motionless object could be enclosed in a transparent cylinder wide enough in diameter to prevent the spider from approaching closer than 2.5 cm. Thus the creeping portion of the response chain could be repeatedly elicited, while the jump was rarely if ever performed since the spider could not get within jumping distance. Tubes of narrower diameter, while not allowing the prey to be touched, did not inhibit jumping.

Figure 5A shows the number of jumps elicited before and after rest periods of either 30 or 60 min. The degree of response recovery is clearly greater for the longer rest. However, even after 60 min, there is an extremely large response decrement compared with the initial testing. The degree of recovery over time decreases after 30 min, and at 120 min only slightly more than 30% of the initial response is recovered (Fig. 5B). After exhaustion of the running approach, recovery is more rapid than in the jumping, yet only 50% response recovery is evidenced 2 hr after testing (Fig. 5C). Following exhaustion of the initial orientation to the prey, the response fails to recover completely after 24 hr (Fig. 5D).

Exhaustion of the jumping response somewhat inhibits the approach, while exhaustion of the approach almost completely inhibits jumping. The recovery time course of the approach is not affected by repeated eliciting of the orientation. Nor does feeding with *Drosophila* between tests affect the

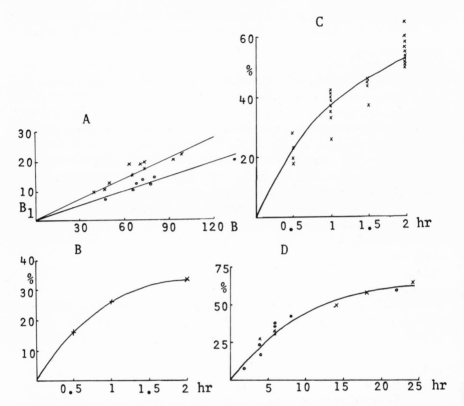

Fig. 5. Recovery from habituation of jumping, running, and orientation in Salticidae (from Precht and Freytag, 1958). A, The number of jumps initially evoked (*B*) is plotted against the responses (B_1) to the same target following either a 1-hr (*X*) or 30-min (*O*) recovery period. B, The time course of spontaneous recovery of jumping is plotted as percent recovery. C, The time course of spontaneous recovery of running plotted as in B. D, The time course of spontaneous recovery of orientation plotted as in B.

recovery of jumping in *Epiblemum*. Feeding *Evarcha* with *Drosophila* and *Epiblemum* with larger flies does retard recovery. (Gardner, 1964, 1965, 1966, has investigated the relationship between hunger levels and the various components of the prey behavior.)

2. *Limulus polyphemus*

The net results of previous works which attempted to assess *Limulus'* learning ability have been largely inconclusive. Experiments were undertaken with the aim of more precisely controlling input and output variables while demonstrating a simple form of learning—habituation. *Limulus* provides a rather advantageous preparation for the study of learning and its physiological basis. It is an extremely large animal with a large nervous system, permitting easy dissection and precise localization of electrodes and lesions. Furthermore, as is typical of the Chelicerata, the nervous system is entirely surrounded by a tough arterial sheath. A cannula may be inserted into this sheath, permitting easy perfusion (Von Burg and Corning, 1970). Corning et al. (1965) have also reported methods for chronic implantation of electrodes in the lateral eye optic nerve, in the ventral cord, and dorsally for EKG recordings, as well as chronic cannulation for perfusion of the circulatory system (pericardial sinus) with drugs or isotopes.

Considerable information has accrued concerning the transducing properties of the lateral eye in *Limulus*, but little is available on the nature and the plasticity of behaviors affected by visual input. Cole (1922) indicated that freshly collected arrivals are positively phototropic but that this response is highly variable. Mere handling of the animal, starvation, and tenure in the laboratory could alter the response. More recent research has shown that unilateral excitation of the lateral eye in a restrained animal elicits a rather definite array of responses (Corning, 1966; Corning and Von Burg, 1968). The telson was observed to swing toward the light source, the flaplike endings at the end of the fifth walking leg closed, and walking movements occurred mainly in contralateral appendages. The flap-closing reflex was always contralateral, whereas with the walking leg movements there was some degree of ipsilateral activation. A response hierarchy was also obtained;

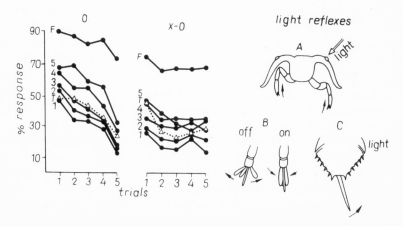

Fig. 6. Habituation of visual reflexes in *Limulus*. A, Cross-section showing relative position of the lateral eyes. The reflex is elicited in contralateral appendages. B, Flap-closing reflex—light off and light on. C, Tail reflex to light stimulation. *O* indicates the average percent response for the reflex in legs 1–5, in addition to the flap-closing and tail reflexes when the original eye was stimulated. *X-O* indicates the average percent response during cross-optic tests (from Corning and Von Burg, 1968).

certain reflexes were more reliably elicited than others. The flap-closing was the most consistently obtained, followed by the fifth, fourth, third, second, and first legs (see Fig. 6). These reflexes showed plasticity. Unilateral excitation of the lateral eye resulted in a decrement in all reflexes (Fig. 6), with the hierarchy persisting throughout habituation. Animals were presented 20–40 habituation trials per day, with each trial consisting of 10 sec of light. Fatigue and sensory adaptation could be ruled out as explanations for the response decrement. Electric shock delivered across the posterior carapace could reliably elicit leg movements for over 400 presentations. Since it took subjects an average of 200 trials to habituate, the motor systems were not fatigued. The carryover of habituation from one day to the next could rule out sensory adaptation. Following habituation, the opposite, unstimulated eye (which had been covered during the trial sessions) was exposed and the animals were tested for cross-optic transfer of the habituation effect. These findings, summarized in Fig. 6, demonstrate a central mechanism for the plasticity.

Visual input produces a transient inhibition of the heart rate, an effect which is mediated by dorsal nerves 7 and 8 emerging from the postesophageal ganglia (Corning and Von Burg, 1968; Corning et al., 1970, 1971). The discovery that the heart rhythm was sensitive to sensory changes led to a series of studies concerned with the pharmacology, electrophysiology, and function of central cardioregulatory units (Corning and Von Burg, 1970; Corning et al., 1970; Lahue and Corning, 1971; Von Burg and Corning, 1967, 1969, 1970). In the course of these studies, the dorsal nerve cardioregulatory units were found to be readily habituated.

Preparation of a crab for a habituation experiment involved fixing it to a board, after which the walking legs were removed to prevent their interference with the recording and stimulating apparatus. The gill books were sectioned medially to expose the ventral cord. Recordings were made with suction electrodes from the dorsal and ventral roots of the first four abdominal ganglia. Ganglion output was measured in the dorsal nerves, which contain cardioregulatory efferents. Tactile stimulation of the gill-book surface was produced by a discrete puff of air. This afferent activity is carried by the ventral nerves.

The response of dorsal nerve units to tactile stimulation of the gill books was a marked increase in rate of firing. When the stimulus was repeatedly applied at a rate of one every 0.73 sec, the initial marked dorsal nerve response decayed rapidly (Lahue and Corning, 1971a). A total of 360 puffs of air were delivered to the animal at this rate. This "acquisition" phase of the experiment was followed by a 4-min rest period in which no stimuli were presented. An additional 360 stimuli were then delivered. This constituted a "4-min retention test." Twelve minutes without stimulation were followed by the "12-min retention test," which consisted of a further 360 puffs of air.

Immediately following each series of stimuli, the stimulated point was given one of three "dishabituation" stimuli: a drop of water, a stronger puff of air, or a touch with a blunt dissecting probe.

During the first 60 air puffs, most units showed a rapid decay in responsivity. This decay was not due to fatigue of the unit, since other stimuli or a stronger puff could restore the responsivity of the unit. Recordings of the sensory nerve units indicated no decrement in ganglion input. In the intact preparation, significant retention (more rapid habituation) was observed for both the 4-min and 12-min retention tests. Occasionally, "off" units were found. Application of the air puff produced a cessation of their spontaneous firing. This effect was also found to habituate and to be retained.

Following this basic demonstration of unit habituation, an attempt was made to assess the relative importance of neural complexity as it affects habituation (Lahue and Corning, 1971b). Four distinct surgical preparations were studied. In the first ("intact") group, recordings were made in a completely intact central nervous system. A "ventral cord" group consisted of animals in which the abdominal ganglia had been isolated from higher influences by lesioning the connectives between the postesophageal ganglia and the first abdominal ganglion. In a third ("isolated ganglion") preparation, recordings were made from a single ganglion isolated by lesioning the connectives immediately anteriorly and posteriorly to one segment. The "one-half isolated ganglion" group consisted of a single isolated ganglion which was split medially into two distinct halves.

In all preparations, the response decrement occurred largely during the first 60 puffs of air. The intact, ventral cord, and isolated ganglion groups did not differ statistically from one another during this period. However, the one-half isolated ganglion group did significantly differ from the intact and ventral cord groups during the same period (see Fig. 7A). Following this initial rapid response decay during the first 60 trials, unit activity in all groups leveled off and remained low for the ensuing 300 stimuli.

During the 4-min retention test, there were significantly fewer spikes recorded in all groups when compared with the same period in the acquisition phase (see Fig. 8A–D). The initial responses were lower and the rates of decay higher in all groups, although the magnitude of this effect was not as great in the isolated ganglion and one-half isolated ganglion groups, both of which were significantly poorer than the other two preparations (see Fig. 7B). As in the acquisition period, the response levels for all groups remained low for the remaining stimuli.

During the 12-min retention test, there was a slight increase in nerve spike activity during the first 60 trials when compared with the 4-min retention test, but this was not significant. However, comparison of the counts obtained during acquisition and those obtained during the first 60 trials of

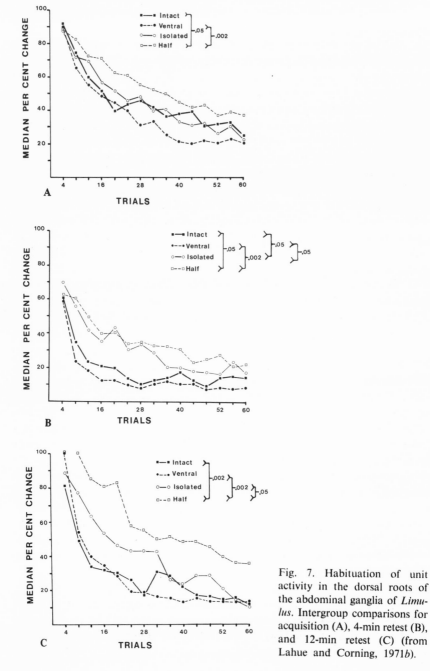

Fig. 7. Habituation of unit activity in the dorsal roots of the abdominal ganglia of *Limulus*. Intergroup comparisons for acquisition (A), 4-min retest (B), and 12-min retest (C) (from Lahue and Corning, 1971b).

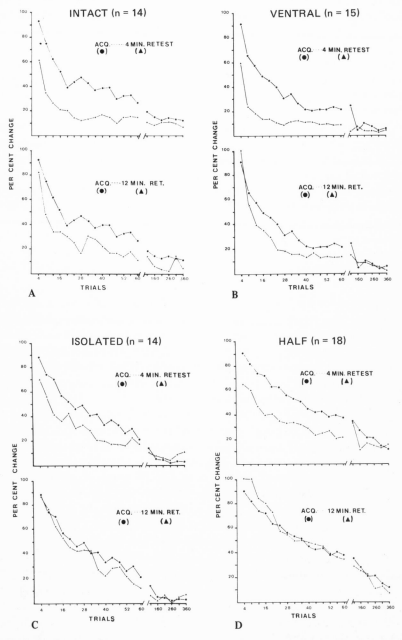

Fig. 8. Intragroup comparisons of acquisition *vs.* 4-min retest and acquisition *vs.* 12-min retest for the same groups and data as plotted in Fig. 7 (from Lahue and Corning, 1971*b*).

the 12-min retest yielded significantly lower scores for the intact and ventral cord groups (see Fig. 8A,B) while the isolated and one-half isolated ganglion groups were returning to the levels observed during the initial acquisition (see Fig. 8C,D and also Fig. 7C).

In summary, all but the one-half isolated ganglion group were equal in terms of the initial acquisition of habituation, but the isolated ganglion and one-half isolated ganglion groups demonstrated less retention during the first 60 trials in both the 4- and 12-min retention tests.

Additional evidence for habituation in *Limulus* central nervous system units has been reported by Snodderly (1969, 1971). Recent studies have elucidated some of the mechanisms involved during habituation in *Limulus* abdominal ganglia. A model of habituation in *Limulus* is discussed with reference to previous models in Lahue and Corning (1973). Extracellular microelectrode recordings in the second optic ganglion showed that units in this region will habituate to repeated light flashes.

B. Conditioning

The Peckhams (1894), working with wolf spiders (*Astia vittata, Phidippus morsitans,* and *Xysticus ferox*), reported the first associative conditioning experiment. Unfortunately, they were more interested in demonstrating color vision than in assessing conditioning capabilities, and few controls were run. They failed to unequivocally demonstrate color vision but did demonstrate clearly that these spiders can establish an association between colored paper placed around their nest site and the nest itself.

Controversy over wolf spiders' memories for their cocoons attracted the attention of many early arachnologists. After laying her eggs, the female wolf spider carries the egg sac for several weeks until the young hatch. Whether the spider will accept a cocoon or a substitute for one which has been re-moved depends on several factors, including the age of the cocoon, time of test-ing, species, and pre-experimental bias of the investigator (Bonnet, 1947).

Berland (1917) confined individual *Nemoscolus laurae* E. Simon to borel tubes, the dimensions of which did not allow tube webs of normal propor-tions to be constructed. A modified web was built, and after removal of the spider to larger quarters the unnaturally shaped web continued to be con-structed for 2 months, never returning to the original form. A similar modi-fication of innate motor patterns was reported by Precht and Freytag (1958). Male salticids induced to courting in tubes, which restricted the normal species-specific leg motions, adapted a new pattern which was retained even after removal from the tube.

Dahl (1884) noted that *Attus arcuatus* Cl. avoided bees and certain beetle species after a few attempts at eating them. Similarly, edible insects normally accepted as food were rejected after a few encounters with turpentine-coated individuals, although other species were still eaten. The negative association was retained for only a few hours. Turnbull (discussed below) reported more enduring positive food associations.

Bays (1962) trained five *Araneus diadematus* to associate dead flies dipped either in sugar or in quinine with the activation of tuning forks at 262 or 523 cps, respectively. Although the initial response in either situation was to bite the fly, by the fifteenth trial the aversive (quinine) fly was discarded without biting—presumably a conditioned response to the tone. After the tones were reversed for four trials, only five trials were needed to reacquire the initial discrimination. When unflavored glass beads were substituted for flies, one bead was always discarded while the other was always bitten. Walcott (1969) replicated this experiment in part with a different species, *Achaearanea tepidariorum*.

While maintaining spiders in the laboratory, previous workers noted that the response of the captive animals to food changed with repeated feeding. Bonnet (1947) has described the behavior of certain spiders which obviously anticipated feeding.

For the sake of verifying such reports and quantifying the nature of the response change, some experiments were performed on a population of *Araneus diadematus* (Lahue, unpublished observations). Specimens were housed individually in inverted clear-plastic drinking cups. All spiders either constructed a miniature web in the cup or left such a number of draglines crisscrossing the interior that they were easily able to move about. The usual maintenance of the animals involves placing a fruit fly under the edge of the cup each day as well as a drop of water at the bottom of the cup. If one turns the cup upright and attempts to feed the spider with a fruit fly held in a pair of forceps, the reaction of 80–90 % of the individuals tested is to run to the bottom of the cup and remain motionless or to run all around the cup avoiding the forceps. However, if the fly is repeatedly put in contact with the mouth region, the spider will eventually bite and eat it. Given one trial daily, a large percentage of the spiders will eventually run to the top of the cup and await the fly. Many even reach out for or jump upon the end of the forceps to obtain the fly.

In this experiment, a positive response was recorded if the spider accepted the fly with a latency of 5 sec or less after the initial presentation. One group of subjects was presented with a fly every day at the same time. Another group received the same treatment but on alternate days. Two control groups were tested in the usual manner on the first and last days of the experiment, but on intervening days flies were placed under the edge of the cup only. One group was fed daily, while the other received a fly on alternate days. Four separate groups were run in each of the control conditions, with a total of 64 spiders fed on alternate days and 65 fed daily. Four groups (64 spiders) were

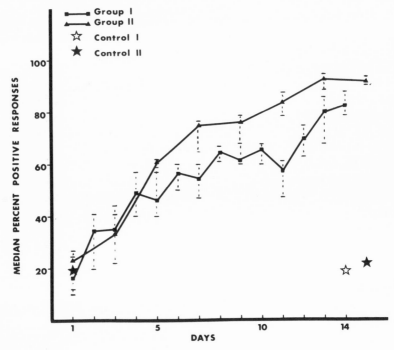

Fig. 9. Acquisition of an approach response to an initially aversive stimulus in *Araneus diadematus*. Group I was trained daily, while group II was trained on alternate days. Control I and control II were fed daily and on alternate days, respectively. Other details in the text (Lahue, unpublished observations).

also run in the daily experimental condition, and three groups (46 spiders) were fed on alternate days.

As can be seen in Fig. 9, the median percentage of animals yielding positive responses on successive days increased in a regular fashion during the course of the experiment. In the group that was treated daily, a significant increase in response was obtained by day 4 in comparison with day 1, while on trials 1 and 2 (i.e., days 1 and 3) the second group already showed significant differences. On the day of training, each group also showed significant differences not only from their response on the initial training day but also from the responses of controls on both initial and final test days. Beginning on day 5, which represented the fifth trial for group 1 and the third trial for group 2, the scores for the two groups differed from one another and this difference was maintained for the remainder of the experiment.

Not only does the initial avoidance reaction elicited in most animals become inhibited very rapidly, but many animals actually demonstrate an approach to the initially aversive stimulus. In addition to recognizing a stimulus situation from day to day, they react adaptively to it.

After demonstrating the remarkable degree of specificity which salticid spiders exhibit in their response to unrewarding stimuli, Drees (1952) investigated behavior which he had observed in nature. *Salticus* will attack ants, but, as these are aversive prey, they are rejected. After jumping on a few ants, the spider ceases to react to them. During the short interval during which it refuses to respond to ants, the spider is also unresponsive to other nonaversive stimuli such as live flies. The responsiveness soon recovers, but after a few more ant attacks the spider again ceases jumping. Following a few such cycles of responsiveness and nonresponsiveness, ants are completely ineffective stimuli for periods of up to a day while the response to flies returns to normal.

Using the shock apparatus pictured in Fig. 3(E), Drees attempted to train spiders to cease responding to a moving two-dimensional target. This apparatus, shown in Fig. 3(C), slid in the experimental box which he used in all his experiments. The immediate response following a single shock experience was to completely cease responding positively to anything. Frequently, a moving target would cause a spider to take flight. After a period of minutes or hours, some responsiveness was recovered, although anything vaguely resembling the form of the training target was avoided. After several negatively reinforced jumps, 46 of 58 subjects ceased approaching the targets and jumping altogether for several days. Of the 58 animals, four showed a specific avoidance of the target associated with shock but would jump to other targets. Eight others would run up to the target but would not jump.

Of the four spiders demonstrating the clearest response, two were trained to a triangle and two to a cross. When other targets were tested (without reinforcement), the number of jumps elicited from the four subjects showed generalization to similar test targets while more dissimilar targets elicited a response.

In analyzing the prey which enter the web of *Linyphia triangularis* under natural conditions, Turnbull (1960) frequently observed that spiders which typically fed on certain species rejected other species which entered the web. Although this fact alone would indicate a real prey preference such as found by Gardner (1965) for jumping spiders, the same spiders were often seen later in the season consistently rejecting the previously acceptable prey and *vice versa*. This suggests that these spiders do have rather specific preferences but that these are not fixed and may be altered during the season. Food preference is probably not innately determined as previously suspected but is, perhaps, learned and adaptive.

Turnbull analyzed the proportion of individuals rejected by *L. triangularis* on the first and subsequent observation days at weekly intervals. The first individuals of most potential prey species entering a web are not as attractive to the spider as individuals entering later. A species which is

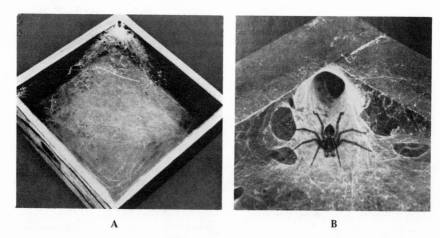

A   B

Fig. 10. A, Wooden box, housing the funnel web of *Agelena labyrinthica*. B, Close-up of the same web showing *Agelena* sitting at the mouth of the funnel-shaped retreat. This is the characteristic position assumed while awaiting prey (from Görner, 1958).

unfamiliar is less likely to be accepted as prey than a species with which the spider has had experience.

Having established a preference for a species, the spider requires "reinforcement" to maintain the preference. If the number of prey of the species which enter the web declines, the spider will establish a new preference for whichever species becomes more abundant. This may be indicative of the relative stability of a learned experience for spiders and suggests some interesting manipulations which could be performed in a controlled study.

Other researchers have described studies which indicate amazing degrees of plasticity in many spider species regarding food selection and predatory behavior (Dabrowska-Prot *et al.*, 1968; Turnbull, 1956, 1964, 1965; Horn, 1969; Luczak, 1970; Kajak, 1965; Haynes and Sisojevic, 1966; Gardner, 1964, 1966). However, in general, such studies merely hint at possible lines of investigation and do not contribute much to the typical analysis of learning phenomena as such.

C. Acquired Orientations

The funnel-web spider *Agelena labyrinthica* awaits prey in the mouth of a densely woven silk funnel (retreat) attached to a flat, nonviscous catching web (see Fig. 10A). The spider very rapidly runs to a vibrating prey entangled in the web, immobilizes it, and returns to the retreat where it eats while waiting for other prey. When *Agelena* is introduced to a wooden box (30 by 30 by 15 cm, as in Fig. 10B) covered with a sheet of glass, it will readily build a web between the walls of the box (Bartels, 1929). The alteration or elimination of optical, gravitational, and fiber-tension cues is easily accomplished, permitting the innate and learned components of web orientation to be tested.

1. The Return to the Retreat

Vibrations in the web stimulate the run to an ensnared prey. Only large (potentially dangerous) intruders are avoided. Inanimate objects are not detected unless vibrated with a tuning fork.

The return from the capture site to the retreat is more complex. The condition of the web as well as its surround is taken into account en route to the prey and integrated with reflex movements. The reflex component of this orientation, while simplifying calculations and speeding the return under normal conditions, is easily overridden by certain experimental manipulations. Any alteration in the sensory cues is likely to cause the spider difficulty in finding the retreat. Figure 11A shows the course taken by a spider returning to the retreat under normal conditions. Figure 11B shows the disorientation caused by switching one light off and another on at the moment that a fly is placed on the web. Similar results are obtained by rotating the web in a windowed room, while web rotation in an optically cueless room has no effect.

Other cues may be altered while in the presence or absence of optical cues. Generally, more severe disorientation results with the exclusion of optical information. Deforming the square box to a rhombus causes disorientation when the retreat is located in an obtuse corner but not when it is in an acute corner (Baltzer, 1930). The disorientation is seen to center along the stretched web diagonal. Measurements of web tension (Holzapfel, 1933a,b), in fact, indicate that in the normal web the retreat lies along the lines of greatest tension.

A reflexive kinesthetic orientation also exists. With small prey, the spider often runs directly to the prey, captures it, and turns 180°. This

Fig. 11. Disorientation of *Agelena*'s return to the retreat caused by altering optical cues (from Bartels, 1929).

"Umdrehreflex" orients the spider properly for a straight return to the retreat. With larger prey, the spider is usually required to make several movements during the capture. A simple 180° turn is not likely to orient the spider toward the retreat in such cases, and the reflex is not seen. The reflex can also be rendered useless by leading the spider in circles around the web with a vibrating tuning fork before allowing it to catch a fly.

The orientation of *Agelena* combines learned as well as innate components. Optical orientations, of necessity, must be learned, while the "Umdrehreflex" is thought to be totally innate. Cues derived from the web itself such as rigidity patterns may be either learned or innate, although experiments in which spiders are transferred to the webs of other spiders seem to support the latter. Learned components are more potent than innate ones. Thus with altered optical cues the spider is often seen to perform the "Umdrehreflex," proceed a short way toward the retreat, then orient optically and head in the opposite direction.

2. Learned Components in the Retreat Orientation

The question may be asked, how rapidly can the spider process enough information to establish an optical orientation? Bartels (1929) found that after he reversed optical cues, *Agelena* required from 26 to 104 min to develop an optical "engram" with which the spider could orient for a return to the retreat after catching prey. In consecutive experiments on two spiders, the time required to learn an orientation increased dramatically over trials when the cues were changed before each trial, indicating the development of a complete state of confusion on the spider's part. Bartels concluded that the memory was a lasting one, since previous learning interfered with later orientations.

Such an interpretation has been criticized in light of recent experiments. Görner (1958) agrees that 30 min is ample time to establish an optical orientation but a disorientation in the spider's return to the retreat occurs only if the cues are changed after the spider arrives at the prey (see Fig. 12A,*a*). Changing the cues before the prey is presented does not disrupt the return (Fig. 12A,*b*). Even if the optical cues are changed only 30 sec before the spider leaves the retreat, changing them again when the spider reaches the prey results in disorientation (Fig. 12A,*c*). If the cues are switched before the spider leaves the retreat after a training period of only 30 sec, no disorientation results (Fig. 12A,*d*). This suggests that the spider has established a new optical orientation in this short period and clearly conflicts with Bartels' findings.

Görner not only concluded that 30 sec is sufficient time to establish an orientation but also hypothesized that there are two sources of information used in the orientation. The first consists of information processed while the

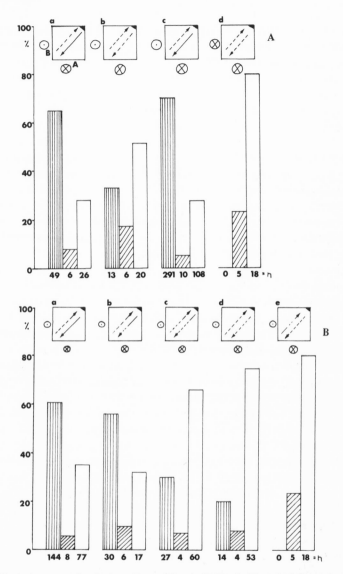

Fig. 12. Optical memory in *Agelena* (from Görner, 1958). A, The vertically striped bars represent the percentage of spiders which showed complete disorientation after the optical cues were altered. The open bars represent the percentage of spiders showing no disorientation. The diagonally striped bars represent spiders demonstrating an intermediate response. Other details are in the text. B, The optical cues were altered at various points en route to the prey: *a*, after the spider reached the prey; *b*, after the spider had traveled three-fourths of the distance to the prey; *c*, after the spider had traveled one-half of the distance to the prey; *d*, after the spider had traveled one-fourth of the distance to the prey; *e*, before the spider left the retreat. The bars are coded as in A.

spider is in the retreat ("Warten-engram"), and the second is influenced by the conditions extant on the way to the prey ("Weg-engram"). With longer training periods in the retreat, the "Warten-engram" predominates, while brief retreat exposure leads to a domination of the retreat orientation by the "Weg-engram." Experiments were undertaken in which the light was switched at various stages of the trip to the prey. Figure 12B,c shows clearly that *Agelena* can relearn its orientation on the second half of the trip to the prey. A further experiment tested the spider's ability to learn an orientation while returning to the retreat. The spider was kept in the dark (dim red light) and then led to a prey. Lamp *A* was turned on as the spider reached the fly and then extinguished just before it reached the retreat. When the spider was immediately lured back on the web (again in the dark) and lamp *B* switched on at the prey site, the spider became disoriented. If on the next run lamp *B* was switched on again, the orientation was good. This was the control trial. On the next trial, lamp *A* was lit and disorientation again resulted.[1]

Similar results have been obtained using orb-weaving spiders. In a position on the web analogous to *Agelena's*, *Araneus diadematus* awaits prey in the center or hub of a more or less vertical, two-dimensional, symmetrical snare (see Fig. 13). While the characteristic head-down position is innately determined, as are the main components of prey-locating behavior, the return to the center of the web (or retreat in some orb-weaving species, see below) is less reflexly controlled and certain manipulations shed light on the innate and learned processes involved. *Araneus* employs two reflex methods to achieve a retreat return. When in the lower half of the web (which usually is slightly less than vertical), dropping from the web on a safety line ("Leit-faden") attached to the hub leaves the spider hanging directly beneath it. A quick run up the thread and the hub position is easily resumed. Although this "Pendel Methode" is probably quite reflexive in nature, the spider must know when to use it. For example, under normal conditions it is never used on the upper half of the web. Using the "Brückenfaden Methode," the spider bites a hole in the web, proceeds to rotate dorsoventrally through the hole, and simply returns straight to the retreat, usually repairing the hole while leaving (Peters, 1931).

If these reflexes were not influenced by external conditions, the spider would always return directly to the retreat. However, as with *Agelena,* turning the web after luring the spider with prey causes disorientation. Often the "Brückenfaden Methode" is performed, but after proceeding a short

[1]It is of interest to note that all experimenters with *Agelena* have acknowledged individual subject variability. Some animals are more optically oriented than others, and it is often possible to test subjects which are completely uninfluenced by optical cues. In the course of their work, most experimenters have selected subjects which orient optically. When studying other orientations, they selected subjects for their weak optical orientations.

Fig. 13. The orb web of *Araneus diadematus*. A 6-inch ruler is included in the picture.

distance toward the retreat, the spider orients according to a learned cue and diverges toward the periphery (both Bartels, 1929, and Moller, 1970, have reported the same behavior in *Agelena*). Altering visual cues disorients normal but not blinded spiders, while elimination of gravitational cues causes both normal and blinded animals to go astray.

3. Retreat Return in Z. x-notata

Recent work has produced more quantitative data concerning the degree of plasticity involved in retreat orientations. Preliminary observations by Le Guelte (1966*a,b*) indicated that *Zygiella x-notata* became disoriented in the retreat run even if the web was turned before the spider was lured from the retreat. However, much less disorientation was exhibited on subsequent trials.

Bartels (1929) observed a similar savings effect in many of his experiments, although the phenomenon was not systematically investigated.

Zygiella's web differs from Araneus' in the presence of a free sector in the upper corner which lacks radii and spirals, and an adjoining retreat which is linked to the hub by a signal thread. When the web is turned 180° (thus altering the spatial orientation of visual cues), Zygiella engages in several periods of "active searching" in the upper corner where the retreat had been, interspersed with motionless intervals (Le Guelte, 1969). Although Z. x-notata builds a web very early in life which differs only in size from every other web it ever builds, there is an apparent learned component in the actual orientation on the web, as evidenced by the fact that experience on the web increases the amount of disorientation caused by turning.

To test the obvious question as to whether true learning (experience) or mere physiological maturation was responsible for such findings, Le Guelte (1969) raised two groups of spiders from hatching. The first group of six spiders was housed in small individual containers in which they were unable to build a web, while a second group was kept on frames and allowed to build a web daily (45 webs each). Individual feeding ensured equivalent growth. At the age of 45 days, the "container spiders" were placed on frames and allowed to build their first web. The following day the webs were reversed, the spiders were lured from their retreats, and their disorientation was measured. The experienced spiders were more disoriented. That is, the time required to locate the retreat was significantly greater. The experienced group required a median of 45 sec return time, while the inexperienced group exhibited a median of 3 sec.

Further experiments were performed using experienced spiders given repeated trials on the reversed web. Seventy-eight 45- to 60-day-old spiders were divided into six groups with intertrial intervals ranging from 2 min to 24 hr. Between trials, the web was returned to its normal position. The six groups showed no significant differences in active searching or motionless times on the first trial. With 24-and 12-hr intertrial intervals, there were no significant differences, comparing the first trials with the second. The same held true for as many as six trials in the 24-hr group. No data were reported beyond the second trial for the other groups. When the interval was 4 hr or less, there were significant decreases in the searching and motionless times on the second trial. With an intertrial interval of 30 min, six trials were required to bring all ten spiders in a group to a criterion of immediate return. Immediate repetition necessitated eight trials to bring a group of 17 to criterion, although the performance of the two groups did not statistically differ. A 24-hr no-trial rest period diminished performance on the first trial in spiders trained to criterion. However, fewer repeated trials were necessary to attain criterion than on the first day, and the difference was significant. By the fourth day of training to criterion each day, the performance on the first trial differed significantly from the initial trial on the first day.

If spiders were tested and then retested 4 hr later without returning the web to its original position after the first trial, there was not only a significantly improved performance on the second trial but the difference was also significant when compared to the second trial of spiders in which the web was returned to the normal position between trials. When the web was reversed 4 hr before the test trial, the return time was less than for spiders with webs reversed at the time of the test. However, the performance was not as good as the second trial of spiders whose webs remained reversed for 4 hr between trials.

In all of the experiments discussed thus far, the web was reversed just prior to luring the spider from the retreat. Information received en route to the prey site may serve the spider in returning. Since it was of interest to determine the role that such information plays in the orientation, another

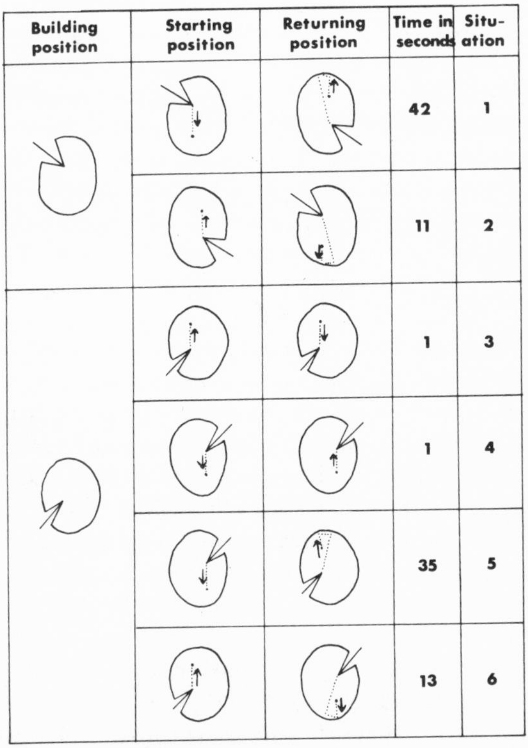

Building position	Starting position	Returning position	Time in seconds	Situation
			42	1
			11	2
			1	3
			1	4
			35	5
			13	6

Fig. 14. Interactions between initial web structure, starting position, and return position and their effect on the time the spider takes to locate the retreat (Le Guelte, 1969b).

series of experiments was performed in which the web position was reversed after the spider was on the web.

A group of 12 spiders was trained to criterion on the usual reversal problem. It was found that on test situations 1 and 2 (see Fig. 14) performance was significantly slower in situation 1. Regardless of previous training, all spiders performed more poorly when the web was reversed after they had left the retreat. Further experiments (situations 5 and 6) demonstrated that when the retreat was positioned above center following the reversal, the spiders performed better than when the opposite situation was tested. As in all experiments (especially noted above in relation to *Agelena*), there was great subject variability. In the experience tests, given the exact same experiences, a wide range of return speeds were exhibited by a group of spiders. This does not exclude certain important conclusions. Le Guelte's experiments have shown that the path taken by the spider in both leaving and returning to the retreat and general experience on the web and inactive experience sitting in the retreat all contribute to the performance of *Zygiella* in retreat orientations.

The last series of experiments seems to indicate a preference for returning in an upper direction, which would correlate well with the normal position of the retreat. Since the structure of the web and the retreat are largely innately determined, this upper orientation may be somewhat innate also, although this point is not very clear in light of the experiments on naive spiders. There is also a tendency for the spider to return in the direction from which it came. This orientation competes with the upper preference and leads to poor performance. *Agelena* shows a similar competition of orientation tendencies which may or may not be classified as innate *vs.* learned.

4. Astronomical Orientation in Wolf Spiders

Certain lycosid spiders which inhabit areas of land bordering water possess an orienting mechanism capable of directing them to shore should they find themselves on water. By synchronizing an internal physiological rhythm with the position of the sun or the polarity of light from the sky, the animal is able to calculate the direction of shore (Papi, 1955a,b, 1959; Papi and Serretti, 1955; Papi *et al.,* 1957). Each segment of the population has an escape direction, which varies depending on the location of shore it inhabits. *Arctosa variana* (L. Koch) is probably the most studied of the species, which also include *A. cinerea* (Fabr.), *A. perita* (Latr.), and *Lycosa fluviatiers* (Blackner) (Papi and Syrjamaki, 1963).

For testing, single spiders were placed in a bowl filled partly with water which was divided into 16 sectors. When placed on the water, the spiders skated rapidly and attempted to escape but crashed into the walls of the bowl. Every collision was called an escape attempt, and the sector in which it occurred was recorded. A measure of the dispersion of *n* escape attempts during a 2-min test resulted from the calculation of the mean escape vector *r* and its length *or* (see, for example, Fig. 15). The bowl sectors were numbered clockwise beginning with north as 0. The direction of *r* was expressed in degrees and the length of *or* could vary from 0 to 1, lower values indicating greater dispersion of escape attempts (for formulas used in the calculations, see Papi and Tongiorgi, 1963).

By procuring eggs laid in the laboratory or unhatched eggs laid in nature and then hatched in the laboratory, young spiders were obtained from seven different females from three different areas. The orientation of newly hatched spiders was not specified, whether reared in the laboratory or taken off the mother's back. Within 30 days, all showed a preference for a north escape regardless of the assumed orientation of the mother. This occurred in young spiders from quite different locations (Papi and Tongiorgi, 1963). As in previous work, this orientation was found to depend on an astronomical orientation which compensates for the apparent movement of the sun. By phase-shifting the animals' internal clock with artificial illumination, it is possible to modify the escape direction (Tongiorgi, 1959).

The initially preferred escape directions of various populations are not innately different, and it has not yet been determined why animals develop the northerly preference. This preference can be modified by artificially moving the "bank" to a new location, and movements on the water are not critical in this orientation change (Papi and Tongiorgi, 1963).

Papi *et al.* (1957) found that although 21 days of captivity under normal illumination did not alter the escape orientation of a group of adult *A. variana,* the orientation did become imprecise. This greater variability of escape attempts was attributed to the effects of testing at differing times of day. A test situation in which no escape is possible is unrewarding and seems to disperse attempts. Several experiments have indicated that in a somewhat more natural setting this phenomenon aids the learning of a new escape orientation (Papi *et al.,* 1957; Papi and Tongiorgi, 1963). Three groups of spiders with a theoretical escape direction of 90° were kept in tanks with water. The escape direction was then changed for two groups (180° and 270°, respectively). After 15 days, the 90° group still oriented toward 90°, while the 270° group made an impressive shift in orientation (see Fig. 15). However, the 180° group did not show an accurate shift in direction. Experiments on *Agelena* indicate that the degree of disparity between an old and a new orientation is inversely related to the ease of reorientation (Moller, 1970). It appears that a small change is not detected as well and the new orientation is imprecise. A large reorientation requirement is clearly detected, and a much more accurate redirection is effected.

Experiments performed in the field also lead to the conclusion that *Arctosa* is able to learn a new escape orientation. Papi and Tongiorgi (1963)

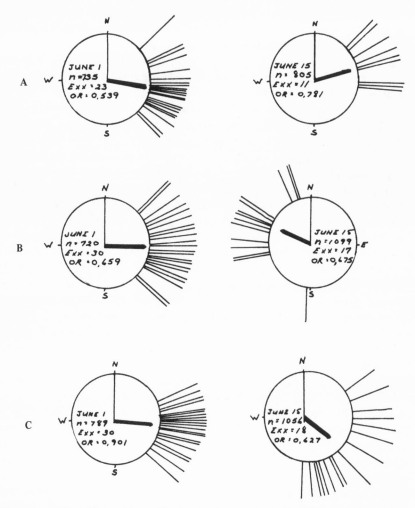

Fig. 15. Learning of a new escape direction in *Arctosa cinerea* (from Papi and Tongiorgi, 1963). The large arrow inside the circles represents the mean escape vector of a group of spiders. Lines outside the circle represent the mean vectors for individual spiders. Each group of spiders was tested twice. Circles on the left represent the initial (control) text. The mean escape direction in each group was approximately 90°. The theoretical escape in A remained 90° but was changed to 270° and 180° in B and C, respectively.

reported certain qualitative aspects of the process. It seems clear that a longer time is required for the learning of a new escape direction by a spider than for the original learning by an inexperienced subject. The time required to learn a new orientation varies markedly between individuals. If training is discontinued before the new orientation is achieved, the spider resumes the old one. However, once a new orientation is learned it often tends to remain stable, although a certain number of subjects gradually return to the old orientation if training is suspended.

Recent work (Linsenmair, 1968) has indicated that scorpions (*Androctonus, Buthus, Buthotus,* and *Scorpio*) learn an orientation by which they position their bodies to make use of cooling wind currents. Such an orientation is dependent on astronomical cues as well as visual landmarks (wolf spiders can also orient to landmarks; Papi and Tongiorgi, 1963).

5. Place Memory on the Catching Web

Another sort of place memory has been investigated in various groups of spiders. Rather than manipulating the spider's orientation to a retreat or safe place, some investigators have observed how well sites on the catching web are remembered.

When presented with more than one prey at a time, orb-weaving spiders such as *Araneus diadematus* bind a captured fly to the hub before retrieving another. Such behavior immediately makes us suspect that the spider remembers not only the presence on the web of an unfinished meal but also its location. Volkelt (1914) observed that spiders continued to search for prey which had been removed from the web. Since the hole left in the web by cutting the fly out may have been a cue for the spider, Baltzer (1923; also Dahl, 1885; Grünbaum, 1927; Müller, 1928) removed a fly which had temporarily been hung from the hub of the web of *Araneus* without disturbing the web or the spider. After being unable to immediately relocate the prey, the spider searched for 26 min, repeatedly returning to the site where the prey was caught. Searching times varied from 4 to 47 min in six other experiments. Other experiments (Peters, 1932) were in agreement with these findings.

Working with *Agelena labyrinthica* and *Tegenaria domestica,* Bartels (1929; Bartels and Baltzer, 1928) demonstrated that the return to a prey site is not necessarily dependent on the spider's actually sensing the presence of the prey. Rather, the spider seems to truly remember the position of the prey on the web and is able to relocate the site using chiefly an optical orientation. Flies were tied with hair to the edge of an experimental box (as described above, Fig. 10), thus preventing the spider from taking them back to the retreat. Normally, the spider alternated intervals of unsuccessfully pulling on the fly with brief returns to the retreat. Eventually, the fly was eaten where it was tied. This indicates a certain degree of plasticity in the animal's behavior, since under normal conditions the prey is always eaten in the retreat. (It would be interesting to know whether or not any evidence of savings would be obtained over repeated trials on the same animal.)

If the fly was removed while *Agelena* was absent, the spider continued to return to the site. Repeated searches for the prey became more diffuse, typically covering ever wider areas of the web. It is not clear whether this represents an adaptive response yielding the spider a better chance of finding a prey which has somehow managed to shift position or whether it simply reflects a gradual waning in the efficiency of an orientation which is unrewarded. In either case it seems clear that the spider still remembers the prey or it would not be on the web searching instead of sitting at the mouth of the retreat.

The total time in which the prey was sought and the number of searches made varies considerably. However, they are certainly much greater than reported by Baltzer (1923). The range observed over several spiders was from five searches in $1\frac{1}{2}$ hr to 19 searches in $3\frac{1}{2}$ hr. Usually, all but the last searches were well oriented to the actual prey site and probably give a fair reflection of the strength of the association which it holds for the spider.

Henton and Crawford (1966) evaluated the ability of tarantula spiders *(Aphonopelma californica)* to establish a position habit and three subsequent response shifts with or without discriminative stimuli. Animals were run in a plexiglas T-maze under three conditions. One group was trained while both arms of the maze were uniformly illuminated; half of these were run under low illumination and half under high illumination. In either condition, half were initially required to turn right and half to turn left. A second group was trained to discriminate between compound cues in the maze. Four subgroups were trained to go to a right–bright, right–dim, left–bright, or left–dim arm. A third group was trained with 90° and 180° polarized light instead of a brightness difference as the discrimination cue. All groups received one trial daily for 20 days. A noxious stimulus (puff of air) was used to motivate the subjects, and running time and number of errors were recorded. After the acquisition period, three successive eight-trial response shifts were given. With respect to running time, the polarized cue group performed significantly better than either of the other groups; all groups decreased running time significantly over trials. No differences were found between groups on any of the reversals, nor did number of errors differ in any of the conditions.

Holzapfel-Meyer (1943) attempted to determine plastic abilities in the behavior of *Agelena* by requiring it to deviate from the straight route to a goal. Various sizes and shapes of obstacles were placed between the spider and a goal, and the subsequent behavior was observed and categorized. The conditions of the experiment were especially appealing because not only could the spider's attempts to reach a prey behind an obstacle be observed but the return to the retreat also required the spider to again navigate a detour. Seventeen animals were tested on 170 trials over a period of 3 years. As a rule, after a "negative" response was made in approaching the prey, the spider's return to the retreat showed no apparent profit from the initial experience with the obstacle. However, spiders responding in a "positive" manner en route to the prey returned to the retreat without encountering the obstacle at all.

D. Conditioning in *Limulus*

Various conditioning paradigms have been employed in attempts to define the limits of plasticity in *Limulus*. Although the results have been equivocal at best, it is not clear whether the limits of the subject or of the experimenter are being taxed.

Makous (1969) attempted to operantly condition joint movements, reinforcing with aerated seawater on the gills or food-flavored seawater on the mouth region. Either stimulus served to generally elevate activity and consequently any response being measured. The effect proved to be transient. Attempts to establish a light–dark discrimination in the task were unrewarding. The author suggests the possibility that daily training sessions were not pursued long enough to produce positive results. Unfortunately, no indication is given of the number of trials which were actually administered before abandoning the experiment.

The earliest attempts at conditioning were those of Smith and Baker (1960) employing a classical conditioning paradigm. Eight dark-adapted male specimens fixed in a holder were presented with a 10-sec light as the CS, the last second of which overlapped with a 1-sec US consisting of a direct-current shock delivered across the posterior carapace. Movement of the telson was recorded in response to these stimuli. Intertrial intervals varied randomly from 100 to 250 sec. Twenty trials per day were administered until the animal reached a criterion of 17 or more responses to the light alone on 3 consecutive days. This was followed by extinction trials until the *S*s returned to their original spontaneous response rates (about eight or less). A light-control group of four was treated the same as the experimental group except for the shock, which was omitted. A shock-sensitization group of six was shocked ten times each day and then given the same treatment as the light-control group.

Electric shock produced a rapid tail movement differing from the conditioned response only in magnitude. By the fourth day, the experimental group exhibited the CR significantly more frequently than the control groups, which did not show an increase over their spontaneous rates. The light-control group demonstrated that the increased tail movement in the experimental group did not result from an increased sensitivity to light alone. Controlling for shock excluded the possibility of pseudoconditioning.

Makous (1969) reported a replication of the work of Smith and Baker with only minor alterations of technique. However, only two animals were tested, and controls were not employed. On the other hand, unpublished data by Corning (reported in Corning and Von Burg, 1968) failed to replicate this work, in that many animals (as many as 70%) could not be trained. Furthermore, the responses elicited in the trainable animals were not as consistent over time as those reported by Smith and Baker.

Further experiments by Makous (1969) tested younger animals with modifications of experimental parameters. In 16,000 trials administered to ten Ss, only 76 positive responses were observed, while 251 spontaneous responses were recorded. The CS apparently served to decrease the rate of response rather than increase it.

Wasserman and Patton (1969) essentially employed the technique of Smith and Baker in an avoidance paradigm where tail movement during the US caused the shock to be omitted. As a modification of previous procedures, a harness was employed which caused the telson to float in the air, thus allowing potentiometric determinations of small movements either up- or downward. In this apparatus, the response to shock was always an upward tail movement, while the avoidance responses were usually downward tail deflections. Smith and Baker noted only upward movements as the conditioned responses, but it is possible that their apparatus did not allow downward movements. Six of the seven animals successfully reached the criterion of nine consecutive avoidance responses following a spontaneous response rate of 35%. One animal required only three reinforcements in 200 trials. However, subsequent data place this experiment and previous work in a questionable light.

The spontaneous response rates as well as the control response frequencies in Smith and Baker's experiments were about 30%, as were those in the experiments of Makous and Wasserman and Patton, although Smith and Baker culled out subjects with high spontaneous response rates. But Wasserman and Patton (1970; Wasserman, 1970, 1972) reported further experiments in which naive *Limulus* responded to light with tail movements on more than 75% of the trials. This response is so reliable that rather accurate light threshold estimations are possible in some subjects. Such variables as time of year and location of the animal supplier can significantly affect which animals yield high spontaneous response rates.[2]

Makous (1969) also attempted to instrumentally condition a locomotor response in *Limulus*. By use of a plastic tray which simulated a shuttle box, juvenile animals were trained to either avoid or approach a 10-sec light at

----

[2]Lahue and Corning (unpublished data) have modified a previous preparation (Corning *et al.*, 1965) by inserting microelectrodes into the various muscles controlling telson movement. The monitoring of muscle unit activity seems to offer a more sensitive means of assessing activation of the telson reflex.

one end of the tray in order to avoid a direct current which was passed through the water in one end. Animals were observed to move from one end of the tray to the other on 20% of the trials. Sixty percent of these responses were positive; i.e., shock was avoided. Since this performance was better than would be expected by chance, a sufficient number of trials could theoretically produce a significant effect. Furthermore, the experimenter observed that on many trials in which the subject failed to avoid, there was an anticipation of the shock. It would seem, then, that an appropriate manipulation of the technique could make this type of conditioning an extremely rewarding one. However, in light of *Limulus'* notorious lack of cooperation in behavioral studies, these require replication.

IV. CONCLUSIONS

Although habituation has been demonstrated throughout the Chelicerata, in no case has there been a systematic investigation of the various levels of habituation in any particular species or even of one level across several species (see Chapter 1; also Corning and Lahue, 1972). Such an analysis is underway in *Limulus* (Lahue and Corning, in preparation). Extension of the experiments with this most primitive chelicerate to the somewhat more advanced Scorpionida may yield an insight into the types of functional adaptations in plasticity mechanisms which were required by the terrestrial mode of life. Although analysis of spider habituation on the electrophysiological level may present some technical problems, due to the generally small size of spiders, certainly these difficulties could be overcome in some of the larger species.

While little is known of the behavior of scorpions and no significant learning experiments have been performed, they do possess some interesting behavior and should provide excellent subjects for future analysis. Although the behavior of *Limulus* seems less varied and a number of behavioral investigations have failed to clearly demonstrate higher forms of learning, new experimental designs should be able to solve the problem. On the other hand, the range of behavior exhibited by spiders is truly staggering. Many of the experiments described above indicate that careful attention paid to the details of an animal's natural life can result in the creation of ingenious laboratory experiments capable of elucidating learning mechanisms underlying behavior. One can only wonder why such a diverse group of animals has received so little attention compared with many other invertebrates. Perhaps the following (surely exaggerated) description of lycosid behavior which appeared in a classic article at the turn of the century has intimidated behavioral experimenters ever since:

When a person passes near one . . . it starts up and gives chase, and will often follow for a distance of thirty or forty yards. . . . Riding at an easy trot over the dry grass, I suddenly observed a spider pursuing me, leaping swiftly along and keeping up with my beast. I aimed a blow with my whip, and the point of the lash struck the ground close to it, when it instantly leaped upon and ran up the lash, and was actually within three or four inches of my hand when I flung the whip from me. (W. H. Hudson as cited in Peckham and Peckham, 1894.)

REFERENCES

Abushama, F. T., 1964, On the behaviour and sensory physiology of the scorpion *Leiurus quinquestriatis* (H. & E.), *Anim. Behav., 12*, 140–153.

Adolph, A. R., 1971, Recording of optic nerve spikes underwater from freely-moving horseshoe crab, *Vision Res., 11*, 979–983.

Baltzer, F., 1923, Beiträge zur Sinnesphysiologie und Psychologie der Webespinnen, *Mitt. Naturforsch. Ges. Bern.*, 163–187.

Baltzer, F., 1930, Über die Orientierung der Trichterspinne *Agelena labyrinthica* (Cl.) nach der Spannung des Netzes, *Rev. Suisse Zool., 37*, 363–369.

Barrows, W. M., 1915, The reactions of an orb-weaving spider, *Epeira scoleptaria* Clerck, to rhythmic vibrations of its web, *Biol. Bull., 29*, 316–332.

Bartels, M., 1929, Sinnesphysiologische und psychologische Untersuchungen an der Trichterspinne *Agelena labyrinthica* (Cl.), *Z. Vergl. Physiol., 10*, 527–593.

Bartels, M., and Baltzer, F., 1928, Über Orientierung und Gedächtnis der Netzspinne *Agelena labyrinthica*, *Rev. Suisse Zool., 35*, 247–258.

Bays, S. M., 1962, Training possibilities of *Araneus diadematus, Experientia, 18*, 423.

Berland, J., 1917, Adaptation de l'instinct chez une Araignée: *Nemoscolus laurae* E. Simon, *Arch. Zool. Exptl., 56*, 134–138.

Berland, L., 1934, Toiles d'araignées à contrepoids, *Nature (Paris), 2930*, 510–512.

Bonnet, P., 1947, L'instinct maternel des Araignées, *Bull. Soc. Hist. Nat. Toulouse, 81*, 185–250.

Bristowe, W. S., 1958, "The World of Spiders," Collins, London.

Cole, W. H., 1922, The effect of laboratory age upon the phototropic reactions of *Limulus, J. Gen. Physiol., 12*, 599–608.

Corning, W. C., 1966, Habituation in *Limulus polyphemus,* Paper presented before Midwestern Psychological Association, Chicago, May 1966.

Corning, W. C., and Lahue, R., 1972, The invertebrate strategy in comparative learning studies, *Am. Zoologist, 12*, 455–469.

Corning, W. C., and Von Burg, R., 1968, Behavioural and neurophysiological studies of *Limulus polyphemus, in* "Invertebrate Neurobiology" (J. Salanki, ed.), Plenum Press, New York.

Corning, W. C., and Von Burg, R., 1907, Neural origin of cardioperiodicities in *Limulus, Can. J. Zool., 48*, 1450–1454.

Corning, W. C., Feinstein, D. A., and Haight, J. R., 1965, Arthropod preparation for behavioral, electrophysiological, and biochemical studies, *Science, 148*, 394–395.

Corning, W. C., Von Burg, R., and Lahue, R., 1970, Endogenous and exogenous influences on *Limulus* heart rhythm, *Am. Zoologist, 10*, 200.

Corning, W. C., Lahue, R., and Von Burg, R., 1971, Supraesophageal ganglia influences on *Limulus* heart rhythm: Confirmatory evidence, *Can. J. Physiol. Pharmacol., 49*, 387–393.

Dabrowska-Prot, E., Luczak, J., and Tarwid, K., 1968, Prey and predator density and their reactions in the process of mosquito reduction by spiders in field experiments, *Ekol. Polska, 16*, 772–819.

Dahl, F., 1884, Das Gehör- und Geruchsorgan der Spinnen, *Arch. Mikroskop. Anat., 24*, 1–10.

Dahl, F., 1885, Versuch einer Darstellung der psychischen Vorgänge in den Spinnen, *Vischr. Wiss. Phil., 9*, 84–103, 162–190.

Drees, O., 1962, Untersuchungen über die angeborenen Verhaltensweisen bei Springspinnen (Salticidae), *Z. Tierpsychol., 9*, 12–207.

Dzimirski, I., 1959, Untersuchungen über Bewegungssehen und Optomotorik bei Springspinnen (Salticidae), *Z. Tierpsychol., 16*, 385–402.

Fage, L., 1949, Mérostomacés, *in* "Traité de Zoologie" (Grassé, P. P., ed.), Vol. 6, pp. 219–262. Masson et Cie, Paris.

Gardner, B. T., 1964, Hunger and sequential responses in the hunting behaviour of salticid spiders, *J. Comp. Physiol. Psychol., 58*, 167–173.

Gardner, B. T., 1965, Observations on three species of *Phidippus* jumping spiders (Araneae: Salticidae), *Psyche, 72*, 133–147.

Gardner, B. T., 1966, Hunger and characteristics of the prey in the hunting behaviour of salticid spiders, *J. Comp. Physiol. Psychol., 62*, 475–478.

Görner, P., 1958, Die optische und kinästhetische Orientierung der Trichterspinne *Agelena labyrinthica* (Cl.), *Z. Vergl. Physiol., 41*, 111–153.

Grünbaum, A. A., 1927, Über das Verhalten der Spinne (*Epeira diademata*) besonders gegenüber vibratorischen Reizen, *Psychol. Forsch., 9*, 275–299.

Haynes, D. L., and Sisojevic, P., 1966, Predatory behavior of *Philodromus rufus* Walckenaer (Araneae: Thomisidae), *Can. Entomol., 98*, 113–133.

Heil, K. H., 1935, Beiträge zur Physiologie und Psychologie der Springspinnen, *Z. Vergl. Physiol., 23*, 1–25.

Henton, W. W., and Crawford, F. T., 1966, The discrimination of polarized light by the tarantula, *Z. Vergl. Physiol., 52*, 26–32.

Holzapfel, M., 1933a, Die Bedeutung der Netzstarrheit für die Orientierung der Trichterspinne *Agelena labyrinthica* (Cl.), *Rev. Suisse Zool., 40*, 247–250.

Holzapfel, M., 1933b, Die nicht-optische Orientierung der Trichterspinne *Agelena labyrinthica* (Cl.), *Z. Vergl. Physiol., 20*, 55–116.

Holzapfel-Meyer, M., 1943, Umwegversuche an der Trichterspinne *Agelena labyrinthica* (Cl)., *Rev. Suisse Zool., 50*, 89–130.

Homann, H., 1928, Beiträge zur Physiologie der Spinnenaugen. I and II, *Z. Vergl. Physiol., 7*, 201–268.

Homann, H., 1931, Beiträge zur Physiologie der Spinnenaugen. III, *Z. Vergl. Physiol., 14*, 40–67.

Homann, H., 1934, Beiträge zur Physiologie der Spinnenaugen. IV, *Z. Vergl. Physiol., 20*, 420–429.

Horn, E., 1969, 24-Hour cycles of locomotor and food activity of *Tetragnatha montana* Simon (Araneae: Tetragnathidae) and *Dolomedes fimbriatus* (Clerck) (Araneae: Pisauridae), *Ekol. Polska, 17*, 533–547.

Kaestner, A., 1950, Reaktion der Hufspinnen (Salticidae) auf unbewegte farblose und farbige Besichtsreize, *Zool. Beiträge, N.F. 1*, 12–50.

Kaestner, A., 1968, "Invertebrate Zoology," Vol. 2, Interscience, New York.

Kajak, A., 1965, An analysis of food relations between the spiders *Araneus cornutus* Clerck and *Araneus quadratus* Clerck and their prey in meadows, *Ekol. Polska, 13*, 717–762.

Lahue, R., and Corning, W. C, 1971a, Habituation in *Limulus* abdominal ganglia, *Biol. Bull., 140*, 427–439.

Lahue, R., and Corning, W. C., 1971*b*, Plasticity in *Limulus* abdominal ganglia: An exercise in paleopsychology, *Can. Psychol., 12* (Suppl. 2), 193–194.

Land, M. F., 1969*a*, Structure of the retinae of the principal eyes of jumping spiders (Salticidae: Dendryphantinae) in relation to visual optics, *J. Exptl. Biol., 51,* 443–470.

Land, M. F., 1969*b*, Movements of the retinae of jumping spiders (Salticidae: Dendryphantinae) in response to visual stimuli, *J. Exptl. Biol., 52,* 471–493.

Land, M. F., 1971, Orientation by jumping spiders in the absence of visual feedback, *J. Exptl. Biol., 54,* 119–139.

Le Guelte, L., 1966*a*, Note préliminaire sur un apprentissage chez *Zygiella x-notata* Cl. à l'état adulte (Araignée, Argiopidae), *Compt. Rend. Acad. Sci. Ser. D, 262,* 689–691.

Le Guelte, L., 1966*b*, Structure de la toile de *Zygiella x-notata* Cl. (Araignée, Argiopidae) et facteurs qui régissent le comportement de l'Araignée pendant la construction de la toile, *Thèse. Publ. Univ. Nancy,* 1–77.

Le Guelte, L., 1969, Learning in spiders, *Am. Zoologist, 9,* 145–152.

Linsenmair, K. E., 1968, Anemomentotaktische Orientierung bei Skorpionen (Chelicerata, Scorpiones), *Z. Vergl. Physiol., 60,* 445–449.

Luczak, J., 1970, Behaviour of spider populations in the presence of mosquitoes, *Ekol. Polska, 18,* 625–634.

Makous, W. L., 1969, Conditioning in the horseshoe crab, *Psychon. Sci., 14,* 4–6.

Millot, J., 1949, Classe des Arachnides. I. Morphologie générale et anatomie interne, *in* "Traité de Zoologie" (Grassé, P. P., ed.) Vol. 6, pp. 263–319, Masson et Cie, Paris.

Millot, J., and Vachon, M., 1949, Scorpions, *in* Traité "de Zoologie" (Grassé, P. P., ed.) Vol. 6, pp. 386–436. Masson et Cie, Paris.

Moller, P., 1970, Die systematischen Abweichungen bei der optischen Richtungsorientierung der Trichterspinne *Agelena labyrinthica, Z. Vergl. Physiol., 66,* 78–106.

Müller, R. T., 1928, Discussion following Bartels and Baltzer, *Rev. Suisse Zool., 35,* 256–258.

Murphy, R. C., 1914, Reactions of the spider *Pholcus phalangioides, N.Y. J. Entomol. Soc., 22,* 173–174.

Palka, J., and Babu, K. S., 1967, Toward the physiological analysis of defensive responses of scorpions, *Z. Vergl. Physiol., 55,* 286–298.

Papi, F., 1955*a*, Astronomische Orientierung bei der Wolfspinne *Arctosa perita* (Latr.), *Z. Vergl. Physiol., 37,* 320–233.

Papi, F., 1955*b*, Ricerche sull'orientamento di *Arctosa perita* (Latr.) (Araneae, Lycosidae), *Pubbl. Staz. Zool. Napoli, 27,* 80–107.

Papi, F., 1959, Sull' orientamento astronomico in specie del gen. *Arctosa* (Araneae, Lycosidae), *Z. Vergl. Physiol., 41,* 481–489.

Papi, F., and Serretti, L., 1955, Sull'existenza di un senso del tempo in *Arctosa perita* (Latr.) (Araneae, Lycosidae), *Atti. Soc. Tosc. Sci. Nat., Mem. (B), 62,* 98–104.

Papi, F., and Syrjamaki, J., 1963, The sun-orientation rhythm of wolf spiders at different latitudes, *Arch. Ital. Biol., 101,* 59–77.

Papi, F., and Tongiorgi, P., 1963, Innate and learned components in the astronomical orientation of wolf spiders, *Ergeb. Biol., 26,* 259–280.

Papi, F., Serretti, L., and Parrini, S., 1957, Nuove ricerche sull'orientamento e il senso del tempo di *Arctosa perita* (Latr.) (Araneae, Lycosidae), *Z. Vergl. Physiol., 39,* 521–561.

Patten, W., and Redenbaugh, W. A., 1900, Studies on *Limulus.* II. The nervous system of *Limulus polyphemus* with observations upon the general anatomy, *J. Morphol., 16,* 91–200.

Peckham, G. W., and Peckham, E. G., 1887, Some observations on the mental power of spiders, *J. Morphol. Physiol., 1,* 383–419.

Peckham, G. W., and Peckham, E. G., 1894, The sense of sight in spiders with some observations on the color sense, *Trans. Wisc. Acad. Sci., 10,* 231–261.

Peters, H., 1931, Die Fanghandlung der Kreuzspinne (*Epeira diademata* Cl.), *Z. Vergl. Physiol., 15,* 693–747.

Peters, H., 1932, Experimente über die Orientierung der Kreuzspinne *Epeira diademata* Cl. im Netz, *Zool. Jahrb., 51,* 239–288.

Precht, H., 1952, Über das angeborene Verhalten von Tieren. Versuche an Springspinnen (Salticidae), *Z. Tierpsychol., 9,* 207–230.

Precht, H., and Freytag, G., 1958, Ermüdung und Hemmung angeborener Verhaltensweisen, *Behaviour, 13,* 143–211.

Raw, F., 1957, Origin of the chelicerates, *J. Paleontol., 31,* 139–192.

Savory, T. H., 1934, Experiments on the tropisms of spiders, *Queksett Microscop. Club., 53(1),* 13–24.

Sharov, A. G., 1966, "Basic Arthropodan Stock," Pergamon Press, London.

Smith, J. C., and Baker, H. D., 1960, Conditioning in the horseshoe crab, *J. Comp. Physiol. Psychol., 52,* 279–281.

Snodderly, D. M., Jr., 1969, "Processing of Visual Inputs by the Ancient Brain of Limulus" (Thesis), Rockefeller University, New York. Available from University Microfilms, Ann Arbor, Michigan.

Snodderly, D. M., Jr., 1971, Processing of visual inputs by brain of *Limulus, J. Neurophysiol. 34,* 588–611.

Störmer, L., 1963, *Gigantoscorpio willsi,* a new scorpion from the lower Carboniferous of Scotland and its associated preying microorganisms, *Skrift. Norsk. Vid.—Akad. Oslo. I. Mat—Nat. Klasse (Ny Ser.), 8,* 1–171.

Thomas, M., 1933, Immobilisation simulatrice et prescience anatomique, *Ann. Bull. Soc. Entomol. Belg., 73,* 260–262.

Tongiorgi, P., 1959, Effects of the reversal of rhythm of nycthemeral illumination on astronomical orientation and diurnal activity in *Arctosa variana* (L. Koch) (Araneae, Lycosidae), *Arch. Ital. Biol., 97,* 251–265.

Turnbull, A. L., 1956, Spider predators of the sprace budworm *Choristoneura fumiferana* (Clem.) at Lillooct, B.C., Canada, *Proc. Eighth Pacif. Sci. Congr.,* 1579–1594.

Turnbull, A. L., 1960, The prey of the spider *Linyphia triangularis* (Clerck) Araneae, Linyphiidae), *Can. J. Zool., 38,* 859–873.

Turnbull, A. L., 1964, The search for prey by a web-building spider *Achaearanea tepidariorum* (C. L. Koch) (Araneae, Theridiidae), *Can. Entomol., 96,* 568–579.

Turnbull, A. L., 1965, Effects of prey abundance on the development of the spider *Agelenopsis potteri* (Blackwall) (Araneae: Agelenidae), *Can. Entomol., 97,* 141–147.

Volkelt, H., 1914, Über die Vorstellungen der Tiere, *in* "Arbeiten zur Entwicklungspsychologie" (Felix Krueger, ed.), p. 1.

Von Burg, R., and Corning, W. C., 1967, Central nervous system regulation of cardiac rhythms in *Limulus polyphemus, Am. Zoologist, 7,* 278.

Von Burg, R., and Corning, W. C., 1969, Cardioinhibitor nerves in *Limulus, Can. J. Zool., 47,* 735–737.

Von Burg, R., and Corning W. C., 1970, Cardioregulatory properties of the abdominal ganglia in *Limulus, Can. J. Physiol. Pharmacol., 48,* 333–341.

Walcott, C., 1969, A spider's vibration receptor: Its anatomy and physiology, *Am. Zoologist, 9,* 133–144.

Walcott, C., and Van der Kloot, W. G., 1959, The physiology of the spider vibration receptor, *J. Exptl. Zool., 141,* 191–244.

Wasserman, G. S., 1970, Reply to Makous, *Psychon. Sci., 19,* 184.

Wasserman, G. S., 1973, Unconditioned response to light in *Limulus:* Mediation by lateral, median, and ventral eye loci. *Vision Res., 13,* 95–105.

Wasserman, G. S., and Patton, D. G., 1969, Avoidance conditioning in *Limulus, Psychon. Sci., 15,* 143.

Wasserman, G. S., and Patton, D. G., 1970, *Limulus* visual threshold obtained from light-elicited unconditioned tail movements. *J. Comp. Physiol. Psychol., 73,* 111–116.

Wen, Y. N., 1928, Experiments on spiders. Can they learn? *Bull. Peking Soc. Nat. Hist., 3,* 9.

Wolbarscht, M. L., and Yeandle, S. S., 1967, Visual processes in the *Limulus* eye, *Ann. Rev. Physiol., 29,* 513–542.

Young, M. R., and Wanless, F., 1967, Observations on the fluorescence and function of spider's eyes, *J. Zool. Lond., 151,* 1–16.

Chapter 7

LEARNING IN CRUSTACEA[1]

Franklin B. Krasne

Department of Psychology
University of California at Los Angeles
Los Angeles, California, USA

I. INTRODUCTION

The Crustacea have been called the "water-breathing insects of the sea" (Schmitt, 1965). Although they have invaded both freshwater and terrestrial habitats, this description gives a reasonable intuitive feeling for many of the general behavioral and morphological characteristics of the class.

A. Evolutionary Relationships

The precise phylogenetic relationships of the Crustacea to other nearby groups are not fully known. It is generally agreed that the Crustacea evolved, as did the rest of the Arthropoda, from a polychaete or polychaete-like ancestor. However, the exact affinities of the Crustacea to the insects and myriopods, to the chelicerates, and to the extinct group of primitive marine arthropods known as the trilobites are a matter for some debate; indeed, there is some question as to whether they all evolved from a common ancestor and therefore are really to be included as the classes of a single phylum. Similarly, it is clear that molluscs and annelids are very closely related to one another, but there is disagreement over whether molluscs evolved from a segmented

[1]The author wishes to thank Alan Barnebey, Jeff Wine, and Joan Bryan for help in preparing this manuscript as well as for the use of their unpublished data. The chapter was prepared with the aid of USPHS Grant No. 2-ROI-NS-8108-04.

polychaete-like ancestor, in which case their origin might be rather similar to that of the Crustacea, or whether they and the annelids diverged from a more distant common ancestor such as a flatworm (Barnes, 1968; Waterman and Chace, 1960). In any case, since it is clear, as will be seen in other chapters, that virtually all the groups to which the Crustacea are allied contain animals that can learn, and since the sensory and motor abilities of Crustacea are in a general way comparable to those of the insects—in many of whose lives learning is well known to play a highly significant role—it would be remarkable indeed if the ability to learn were not fairly well developed among the Crustacea.

B. Early Demonstrations of Learning

The first attempt to look for learning was made by von Bethe (1898) in the crab *Carcinus*. He obtained negative results but did not really study the matter very extensively. A few years later, Yerkes (Yerkes, 1902; Yerkes and Huggins, 1903) presented convincing evidence that both crabs and crayfish could learn. We shall take one of the experiments with a single crayfish (from Yerkes and Huggins, 1903) as an example. The animal was taken out of the water and placed in a simple two-choice box in which one choice led out of the maze and back to water while the other was locked; the animal was apparently allowed to wander until it escaped, and when it did so it was allowed 3 min to rest and then was run again. It was run in this way for ten trials each day. Between one trial and the next, the maze was "thoroughly washed out" to prevent the accumulation of olfactory cues from the animal himself.

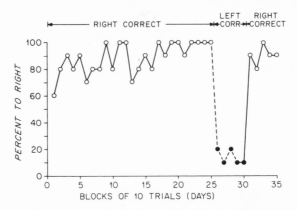

Fig. 1. Position response learning in the crayfish *Cambarus* (data from Yerkes and Huggins, 1903).

It is abundantly clear from the results plotted in Fig. 1 that both original learning and reversal of a previously established habit were rapid in this situation.

Since the time of these early experiments, the fact that many crustaceans can learn has been demonstrated over and over again. Given this, and the fact that it would, as argued above, be remarkable if Crustacea could *not* learn, we will gain little here by a comprehensive recital of experiments whose purpose was merely to prove that this or that crustacean could learn. Nor will we seriously profit by critiques of the methodology of such experiments, although valid criticism can often be raised.

C. Reasons for Studying Crustacean Learning

Other matters should attract our attention in a group such as the Crustacea, whose general grade of organization is clearly sufficient to permit learning and yet whose styles of life are sufficiently diverse to make very different demands on the capacity for modification of behavior.

Great variability within a common pattern of organization in principle makes it possible to dissect out by comparative analysis the relationships among what an animal can learn, the organization of its nervous system, and the demands for learning made on it in life. Except in a few cases, we currently do not understand from either a historical, adaptive, or physiological point of view why some things are learned while others are not. When an animal cannot learn something, is it because the machinery for learning is not there or because it is suppressed? The Crustacea provide superb and almost untapped material for investigating such matters.

As animals that can unquestionably learn but whose attainments in this domain are clearly less developed than our own, the Crustacea also provide excellent material for distinguishing the relatively universal, essential, and primitive aspects of the learning process from more evolved, complex, and derived ones. We might ask, for example, whether slow extinction after partial reward and the capacity for learning by secondary reinforcements (or "token rewards") are fundamental phenomena or are dependent on "higher mental" processes. Or we might ask whether animals that show a capacity to learn in some strict sense of the word will always show a capacity for long-term memory as well. Such questions are readily addressed by the study of groups such as the Crustacea.

Finally, as will be discussed further in later sections, crustacean nervous systems have a variety of features which make them especially valuable for the physiological study of learning.

II. CHARACTERISTICS OF THE GROUP GERMANE TO LEARNING

A. General Characteristics

We have seen that the Crustacea are believed to have evolved from a polychaete-like ancestor. They brought with them a segmented body plan and a well-defined, segmented central nervous system with a considerable degree of cephalization. Their great departures from the annelid plan are the result of the thickening and hardening of their body cuticle. This led from the fleshy parapodia of the polychaetes to stout, jointed limbs whose muscles are attached in such a way as to permit strong and precise manipulation of objects in the environment. The probably primitive (see discussion in Barnes, 1968; Waterman and Chace, 1960; Borradaile *et al.,* 1958) biramous limbs have become differentiated in the various body segments of the various members of the group into highly effective tools for swimming, walking, grasping, biting, grinding, developing water currents, and filtering food-laden water (see Fig. 2).

Effective control of these complex jointed limbs in turn made new demands on the central nervous system and presumably led to an increase in its

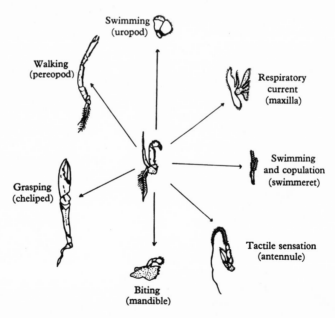

Fig. 2. Adaptive radiation of the crustacean appendage. The different limbs of the crayfish *Astacus* may be regarded as derived from a central type such as the third maxilliped (from Borradaile *et al.,* 1958).

size and complexity. Adaptive use of complex manipulative abilities went hand in hand with elaboration of articulate sense organs, leading, for example, to the group's sophisticated compound eyes. Of course, these improvements in sensory–motor capacity at once permitted, and were selected, by ways of life which used and depended on higher-order abilities to perceive and handle the environment. Thus behavior and the nervous system generally increased in their elaborateness, and, of prime concern for our present purposes, learning became a potentially much more valuable aid to living than it had been in animals with lesser sensory–motor abilities and simpler styles of life.

B. Ways of Life and Learning

Of course, Crustacea are by no means equally developed along the lines of complex sensorimotor equipment. Indeed, one of the greatest unexploited advantages for the comparative analysis of learning abilities is the great diversity of life styles encountered in this one class. Thus there are forms that are specialized for walking or running on the ocean bottom or shore, animals that drift in the plankton, powerful and active swimmers, parasitic forms, and sessile, rooted members of the group. There are animals that are forever adrift in the open sea, animals that build burrows or live in fixed crevices or caves of rocky coasts, and animals that live in the intertidal zone and must deal with its rapidly changing conditions. There are animals that filter-feed or consume mud or silt for what food it contains, animals that are highly particular about what they eat and others that have wide latitudes of taste, animals that graze on seaweed beds or on vegetation which grows on rocks, animals that scavenge decaying plant or animal debris, animals that lie hidden in wait for appropriate living prey which they grab and consume, animals that chase and catch living prey, and animals that dig for buried food.

Thus there are very wide variations in ways of life and therefore in the kind and elaborateness of the sensory, motor, and nervous equipment required for survival. One's expectations as to a particular species' complexity of behavior, "intelligence," ability to learn, and so on, will of course depend on its way of life. Table I, which gives a rough characterization of many (but not all) of the numerous crustacean groups, also attempts to give some feeling as to how the animals live. A great deal can be guessed from the information given in such a table.

We would not expect a sessile, armored, filter-feeding animal such as a barnacle to have much use for the visual acuity provided by compound eyes, and it does indeed lack these, having appropriately reduced the relevant portions of its brain (Bullock and Horridge, 1965). On the other hand, we would expect excellent visual abilities and indications of advanced neural

Table I. General Characteristics of Selected Crustacean Groups[a]

Subclass / Order	Illustration (Fig. 3)	Informal characterization and familiar types	Habitat and mode of locomotion	Food and manner of feeding	Specialized association areas of most rostral division of brain (protocerebrum)	Eyes of adult
1. Cephalocardia	A	Blind, colorless, grublike animals; most primitive Crustacea	Silt, mud, or muddy sand	Organic matter of silt or mud	?	None
2. Branchiopoda	B	Foot breathers; most primitive except for above; orders have little else in common	Planktonic or benthic; see below	Filter feeders on detritus or plankton	See below	Compound; nauplius eye usually persists in adult
3. Anostraca		Fairy shrimps; small, elongate, many-segmented body without a carapace (e.g., *Artemia*, the brine shrimp)	Evanescent fresh waters; swim upside down	See above	Median and lateral groups of globuli cells in *Artemia*	See above
4. Notostraca	C	Tadpole shrimps; small; huge, shieldlike carapace; look like little horseshoe crabs	Evanescent fresh waters; crawl over bottom; can also swim	See above	Median and lateral groups of globuli cells in *Artemia*	See above
5. Conchostraca	D	Clam shrimps; small; completely enclosed in bivalved, closable, clamlike shell	Evanescent fresh waters; plow through mud or crawl on bottom	See above	?	See above

6. Cladocera		Water fleas; small with bivalved shells from which head sticks out; apical spine (e.g., *Daphnia*)	Ocean, lakes, and streams; swim up in jerks by downstroke of antennae and float down	See above	?	See above
7. Ostracoda	E	Mussel shrimps; small with hinged, closable, bivalved shells; antennae may protrude through notch	Marine and freshwater bottoms; scurry over or plow through mud and silt; antennae used for propulsion	Filter feeders	Reduced dorsal group of globuli cells	Compound; nauplius eye persists in adult
8. Copepoda	F	Small pear-shaped, distinctly segmented body; often with spectacular caudal rami (e.g., *Calanus*, *Cyclops*)	Mostly marine; planktonic forms swim by oarlike beating of legs, antenules extended to slow sinking; gliding motion due to feeding current; bottom-livers crawl or burrow with thoracic limbs	Planktonic forms filter-feed but may supplement diet by predation; bottom-livers predacious or graze on detritus	None	Nauplius eye persists in adult; no compound eye
9. Cirripedia		Barnacles	Marine in both intertidal zone and deep water		None	Nauplius eye only
10. Thoracica	G	Acorn barnacles (sessile) (e.g., *Balanus*)	Sedentary on rocks, shells, plants, timber, etc.; commensal on fish and other animals; commensal forms may be host specific; stalks of goose barnacles may be highly mobile	Filter feeders; scoop or rake water; largish prey may be explicitly captured	See above	See above
	H	Goose barnacles (stalked) (e.g., *Lepas*)				
11. Acrothoracica		Naked boring barnacles	Mollusk shells and corals	Filter feeders	See above	See above

Table I (*Continued*)

Subclass / Order	Illustration (Fig.3)	Informal characterization and familiar types	Habitat and mode of locomotion	Food and manner of feeding	Specialized association areas of most rostral division of brain (protocerebrum)	Eyes of adult
12. Rhizocephala and Ascothoracica		Naked parasitic barnacles	Parasitic on decapods, tunicates, echinoderms, and soft corals	Parasitic	See above	See above
13. Malacostraca		Subclass containing three-fourths of known species and all of the larger, better-known, and edible forms	See below	See below	See below	See below
14. Nebaliacea		In many ways reminiscent of branchiopods	Intertidal; swim	Filter feeders; some forms live especially in places foul with organic remains; may stir up bottom to make food available to feeding current	Hemiellipsoid body[b] in eyestalk but no globuli cells	Compound
15. Stomatopoda	I	Mantis shrimps; named after the praying mantis for their raptorial feeding habits; $1\frac{1}{2}$ inch to 1 ft in length; the Milnes believe them	Live in caves or burrows which they plug with their tail fan; swim using pleopods as oars and steer with antennal scales	Predatory, raptorial feeders; may lie in wait for or actively hunt small fish,	Globuli cells near protocerebral bridge and in hemiellipsoid	Compound; nauplius eye persists but small in adult

				crustaceans, and other invertebrates	body[b] of eyestalks; olfactory lobes of deutocerebrum also contain globuli cells	
		to be the most intelligent of the crustaceans (Buchsbaum and Milne, 1960; Milne and Milne, 1961) (e.g., *Squilla*)				
16.		Mysidacea	Opossum shrimps; named for ventral brood pouch, which is also seen in next four orders; look like little shrimp (e.g., *Mysis*)	Include swimming forms and benthic crawling ones	Filter feeders	Hemiellipsoid body[b] and probably also deutocerebrum contain globuli cells	Compound
17.	J	Cumacea	Small Crustacea with relatively huge head and thorax and with brood pouch	Marine; deep intertidal zone; live partially buried in sand or mud; when get too deeply buried by deposits from exhalent current, swim away and dig-in in a new place	Filter feeders	None known	Compound and nauplius in some cases, but many lack eyes
18.		Tanaidacea	Small Crustacea intermediate in characteristics between Cumacea and Isopoda	Marine; intertidal; live partially buried in mud or construct tubes	Filter feeders but may supplement diet by raptorial feeding	No information	Compound in some cases, but many lack eyes
19.	K	Isopoda	Crustacea with dorsoventrally flattened bodies; 5–15 mm long; dorsal exoskeletal plates	Mostly marine but include freshwater and terrestrial forms; most crawl at least part of	Mostly omnivorous scavengers; some carnivorous or	None known	Compound

Table I (*Continued*)

Subclass / Order	Illustration (Fig.3)	Informal characterization and familiar types	Habitat and mode of locomotion	Food and manner of feeding	Specialized association areas of most rostral division of brain (protocerebrum)	Eyes of adult
		tend to project laterally; brood pouch; marine forms: e.g., *Asellus*; terrestrial forms: called woodlice or pill or sow bugs; e.g., *Armadillidium* and *Porcellio*	time; some are very effective runners; aquatic forms can usually swim; many burrow or excavate; terrestrial forms roll up in a ball for protection or to resist desiccation	herbivorous		
20. Amphipoda	L,M	Shrimplike in appearance; similar to isopods in many ways but flattened laterally rather than dorsoventrally; have brood pouch (e.g., *Gamarus* or *Talirus*, the beach fleas); the caprellid amphipods (skeleton shrimps) mimic in color, shape, and posture the marine growths on which they live (e.g., *Caprella*)	Mostly marine; some semiterrestrial, some free-swimming, some bottom-living, all can swim, tend to leave substratum with jump; some adapted with grasping claws for climbing on seaweed; some burrow or construct tubes	Mostly detritus feeders or scavengers; some dig or stir up bottom and then filter it; a few are predators that lie in ambush	None	Compound

21.	Euphausiacea		Shrimplike; about 1 inch long	Marine, free-swimming	Filter feeders	No information	Compound; nauplius eye persists in adult
22.	Decapoda	N,O	Include the shrimps, lobsters, crayfish, and crabs	Mostly marine but some freshwater and terrestrial forms; tend to reside in excavated burrows, natural crevices, constructed tubes, or other natural containers such as the gastropod shells of hermit crabs; shrimps are adapted primarily for swimming; remainder are adapted primarily for walking, although some crabs have become secondarily adapted for swimming; the swimming crabs are the most powerful of crustacean swimmers	Predators and scavengers; swimming crabs can chase and catch rapidly moving prey; many forms supplement their diet by filtering detritus	Hemiellipsoid body[b] in eyestalk usually (but not in some crayfish) in association with globuli cells; olfactory lobes of deutocerebrum also contain globuli cells	Compound; sometimes nauplius eye persists in adult

[a]Information compiled from various sources but especially Barnes (1968), Borradaile et al. (1958), and Bullock and Horridge (1965). Only the most common characteristics are included; exceptions are often omitted.
[b]Hemiellipsoid bodies are thought to be equivalent to corpora pedunculata of insects.

development in very active predatory or scavenging animals such as stomato-pods and crabs, and such animals do indeed possess compound eyes, well-developed visual centers, and what are suspected of being specialized associa-tion areas marked by "globuli cell clusters" in the most rostral parts of their brains (Bullock and Horridge, 1965; and see below).

On the face of it, we would anticipate that an animal such as the lobster, living in a fixed burrow from which it emerges at night to forage, might well learn something about the topography of its home range and the locus of its burrow. We might therefore anticipate that it would be capable of learning its way through mazes in the laboratory. On the other hand, we would not be surprised at the absence of such an ability in a small planktonic creature such as a copepod, which is forever wafted here and there by water currents and knows as objects in its environment only its fellow planktonic creatures whose position relative to one another is ever changing. However, one might expect such an animal to learn what is good to eat and what is not, what is easy to catch (if it is partially predatory, as some are) and what is not, and so on.

While we can make guesses about learning abilities from information about an animal's life style, this does raise a fundamental question which comparative analysis can help to answer. To what extent is learning ability a general property of a nervous system and to what extent is it a highly specific skill? Can an animal that can learn to recognize food necessarily learn to recognize a similar sort of object as an enemy? Will an animal that can learn to approach a given sort of object as food necessarily be able to learn to go to a fixed place to be fed, and so on? Are animals designed to *learn* or merely to learn this and that? Psychologists have usually assumed that learning ability is a general capacity. However, observations on insects (von Frisch, 1953), Crustacea (Bock, 1942), and even rats (Garcia *et al.,* 1968) suggest that this may not be true. We shall discuss this further later.

III. FEATURES OF THE NERVOUS SYSTEM RELEVANT TO LEARNING

A. General Features

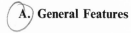

The general features of crustacean nervous systems are similar to those of the polychaetes. Most of the nervous system lies ventrally. In its most primitive condition, its form is that of a ladder with fibers of passage running in the rungs and side rails and with ganglia at the points of intersection. As in the annelids, neuron somas lie at the periphery of their ganglia and con-tribute branching unipolar processes to a central neuropile region in which synapses occur and the information-processing activities of the nervous system are carried out.

The cell bodies of Crustacea do not have synapses and, except for motor neurons, seem not to be invaded to any appreciable degree by electrical events (Takeda and Kennedy, 1964; Remler *et al.,* 1968). This is a particularly interesting situation, because it means that interactions between the nuclear genome and presynaptic endings, post-synaptic receptor sites, and spike-initiating zones will all depend upon relatively time-consuming transport processes. This feature could be important for evaluating various theories about the cell biology of learning and memory, a point to which we will return later.

In all but the most primitive groups, the basic ladder structure has been condensed laterally and rostro-caudally. Thus the rungs of the primitive ladder are often reduced to intraganglionic commissures, and fusions of successive segmental ganglia are common. The most rostral ganglia are enlarged, fused, and dorsally situated to form a supraesophageal ganglion or "brain" which receives the afferent nerves of the special senses as well as sensory information from the whole of the body in an abstracted form (Bullock and Horridge, 1965).

B. The Brain

On the *average,* the brain of Crustacea is more complex than that of annelids in the sense that it has more neurons, more discrete clusters of cell bodies, and more distinguishable fiber masses (neuropile regions). Crustacean brains also contain, more frequently than do those of annelids, particularly fine knots of neuropile (glomeruli) and tight clusters of small, cytoplasm-poor cell bodies (globuli cell masses) which feed neuropile regions of exceptionally smooth, fine-grained texture (Bullock and Horridge, 1965). Such differentiated structures are generally considered as signs of neural advancement. The latter sort, in particular, often receive input from neuropile regions subserving several different senses such as olfaction and vision and are therefore considered to be association areas. The most impressive of these, the so-called corpora pedunculata, have generally been considered, especially in insects, as the seat of the animals' most complex integrative activities; however, the evidence for such assertions is very incomplete and in some cases somewhat contradictory.

While it is true that crustacean brains tend to be more complex in the above respects than those of their polychaete ancestors, there is considerable overlap, with the brains of some errant polychaetes being considerably more complex than those of some Crustacea. Nevertheless, the most advanced arthropod brains are more complex than the most advanced annelid ones, and such complexity generally correlates with sophistication of behavior (Bullock and Horridge, 1965).

We may anticipate that the brain, with its association neuropiles and regions devoted to analysis of special senses, will be important for learning, just as the cortex of mammals is probably involved in the bulk of mammalian learning. However, in the same sense we would not expect the highest centers of invertebrates, any more than of the more encephalized vertebrates, to be essential for all learning and memory.

It is sometimes said that the brain of arthropods in general, and of Crustacea in particular, molds behavior primarily by inhibitory modulation of more or less autonomous segmental reflexes. While it is true that there are some signs of release of function when the brain is removed from a crab or crayfish, as there also are in an insect or an annelid, it is also true that the brain is involved in initiating and controlling a great deal of relatively complex behavior (Bullock and Horridge, 1965). To think of it as operating simply by inhibiting local reflexes is almost certainly a misleading over-simplification. Similarly, the subesophageal ganglion has some tendency to activate or arouse Crustacea, but here too it would almost surely be an oversimplification permitted only by our ignorance to assert that this is *the* thing that the subesophageal ganglion does.

C. Neuron Numbers

There is general agreement (Bullock and Horridge, 1965) that the number of neurons of crustacean nervous systems, as of arthropods generally, is remarkably low for the feats of integration which they can perform. However, only one comprehensive neuron count—on the nervous system of a crayfish—has actually been carried out (Wiersma, 1957).

There are some 70,000–80,000 neuron somas in the brain of a crayfish, of which about 60,000–70,000 are parts of globuli cell clusters; in the subesophageal ganglia there are about 7000 somas, of which 1000 are in globuli masses; in middle thoracic ganglia, which control the animal's manipulative appendages, there are some 2000–3000, none of them globuli cells; in the abdominal ganglia, which do not control appendages that engage in complex manipulation, there are some 500, none of them globuli cells. For comparison, the brain of an earthworm (not a maximally developed annelid brain) contains some 10,000 neurons, while a midbody ganglion contains about 1000 (Bullock and Horridge, 1965). All these numbers are of course minute in comparison to mammalian neuron counts.

D. Ways of Economizing on Neurons

There appear to be several principles of organization which result in a parsimonious usage of neurons in Crustacea.

The first is by abstracting and discarding a great deal of the information in most sensory channels at an early stage of processing and by subsequently dealing with it in a highly reduced form (Wiersma, 1961; Horridge, 1968; Bullock and Horridge, 1965). An exaggeration of a real example will illustrate the point. Suppose the second left walking leg of a crayfish were tickled near its end. The only information about this that the brain gets might well be the firing of neuron A, which responds when a tactile hair anywhere on the most distal part of any leg on the left side of the body is touched, and the firing of neuron B, which responds whenever any sort of stimulus is applied anywhere on the stimulated segment of the body. The exact locus of the stimulus on the leg might be lost entirely to the brain, and even which leg was affected might be determinable only by a process of reconstruction which the brain may or may not be capable of achieving. In fact, one interesting use of learning abilities in animals where sensory channels to the brain are very well known, as they are in the crayfish, will be to take advantage of training techniques to find out to what extent the brain can put such information back together again.

A second technique which appears to result in economy is the use of temporal rather than spatial codes of intensity. Thus, whereas the intensity of contraction of a vertebrate muscle is controlled to a significant degree by the number of motor neurons to the muscle that are active, the intensity of contraction of an arthropod muscle is controlled by the frequency of nerve impulses in a very small number (usually one or two) of motor neurons which feed all the fibers of the muscle (Bullock and Horridge, 1965). If such temporal coding is relied on at all levels of the nervous system, a great economy of neuron numbers could obviously be effected. The difficulty of the method, of course, is that time is required to assess the strength of a given message or directive, so that any complex integration involving serial processing will necessarily take a rather long time. Thus temporal coding is probably not the way to construct a highly intelligent nervous system, but it will result in a marked conservation of neuron numbers.

A third conservation mechanism is the use of one neuron as many. Vertebrate neurons are commonly pictured as being neatly divided into a receptive portion (soma and dendrites) which receives synapses, whose post-synaptic effects spread passively to the axon hillock of the neuron, and an efferent portion (axons and their branches and terminations), in which spikes once initiated spread throughout actively and with certainty. Invertebrate unipolar neurons are functionally much more complex. A given neuron may have dendritic regions in several ganglia, each with its own "axon hillock" or spike-initiating zone. Furthermore, many of the branches of the extensive arbors within ganglia produce spikes in response to local inputs, and such spikes do not necessarily invade the whole tree or main axon but instead may block at a variety of regions of low safety factor such as branch points (Maynard, 1955; Takeda and Kennedy, 1964, 1965). Thus in the invertebrate neuron the weighting of inputs that leads to a decision on whether to fire occurs at many points rather than only at a single axon hillock. It is even possible, at least in principle, that two or more parts of a branching system which are separated by a region of very low safety factor may each have their own afferent and efferent synapses and, except for trophic influences and metabolic controls, may function, in effect, as entirely separate neurons (Hughes, 1965); whether this really happens is not known. However, if it does, soma counts are a poor way of estimating nervous complexity.

The final known mechanism for getting by with small numbers of neurons is the use of what seems to be a rather rigorously hierarchical system of motor control. Basic to this system are so-called command neurons, which have been studied primarily in the crayfish and lobster (Wiersma, 1961; Atwood and Wiersma, 1967; Kennedy, Evoy, and Hanawatt, 1966). These are single units which when stimulated produce a unique item from the animal's total repertoire of coordinated acts. Thus one such command fiber brings about rapid flexion of the abdomen in all its segments, flexion of the telson, forward thrusting of legs, claws, and antennae, and contractions of muscles about the eyes; the net effect of all this is to cause the animal to dart backward through the water. Another fiber causes a crayfish to lift the front part of its body and raise its claws in the typical posture of threat or defense. We shall discuss further examples later. The point, however, is that the remainder of the nervous system need not worry in detail about the formulation of patterns of movement. All it needs to do is play upon the appropriate command fibers at the desired times.

E. Advantage of Conservation of Neurons for Studying Learning

The above principles of construction have very important consequences for the usefulness of Crustacea as suitable subjects for the physiological study of learning. For to a certain extent it appears that in the Crustacea we may not have to solve problems of perception and motor control before getting at the problem of learning. More particularly, it may be possible to think of the points at which various neurons converge on command neurons (or perhaps upon slightly earlier tiers of neurons) as the critical foci of behavioral decisions. One might then expect that it would be precisely at such decision points that the modulations of excitability which constitute learning would occur or at least be applied. If such thinking is valid, the search for the engram would be greatly simplified.

IV. ROLES OF LEARNING IN LIFE

In order to study the learning capacities of a group intelligently and, as discussed above, to appreciate in full measure the significance of such capacities as are found, it is necessary to understand the uses to which learning is put in the life of an animal. The purpose of this section, then, is to survey some of these uses. In so doing, we will not usually be particularly careful about just exactly *what* is being learned or the kind of learning (habituation, associative learning, etc.) that is involved. Rather, we shall be trying to get a feeling for why it is adaptive for Crustacea to learn (in a very broad sense of that word). For this purpose, we shall define learning, broadly, as any change in overt behavior or central state which is a consequence of experience and persists in time.

There are several previous reviews of the material discussed in the present section and Section V (Thorpe, 1956; Schöne, 1961a, 1964; Wiersma, 1970). The review of Schöne (1961a) is a particularly useful review of complex behavior, generally, in Crustacea.

A. Terrain and Place Learning

For Crustacea that move about under their own control in a relatively fixed environment, such as strong swimmers and intertidal, terrestrial, and bottom-living forms do, we may expect that there would be clear advantages in being able to remain subjectively oriented relative to places of interest.

A great many Crustacea live and find shelter in burrows, holes, caves, crevices, etc. In many cases, these are abundant or easily constructed, and their use is on a one-time basis. But in some instances specific shelters are maintained and, indeed, defended over a period of time.

The rock lobster *Panulirus interruptus* (Lindberg, 1955) resides during the day in a variety of natural shelters. These are not always easy to come by, especially for larger individuals. At night, the animals leave their shelters to forage on the bottom hundreds of yards or more from their homes, and we would expect that they might return to these abodes. Although there has been little investigation of this, individuals have been seen to maintain a single residence for weeks or even years. While we could in principle argue that perhaps individualized chemical trails are laid down and followed or that homes are found by trial and error and recognized by comparison of

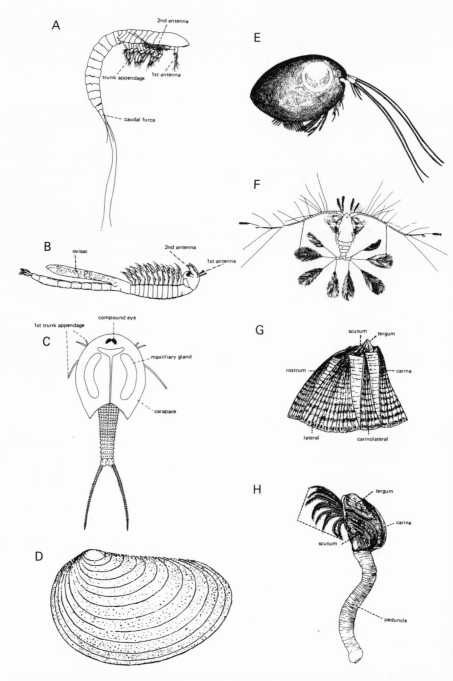

Fig. 3. A range of crustacean types A, A cephalocardian, *Hutchinsonella macracantha*.
B, A fairy shrimp, *Branchinecta*. C, A tadpole shrimp, *Apus*. D, A clam shrimp, *Estheria*.
E, A mussel shrimp, *Cythereis*. F, A marine copepod, *Calocalanus pavo*. G, An acorn
barnacle, *Balanus*. H, A goose barnacle, *Lepas hillii*. I, A mantis shrimp, *Squilla mantis*.

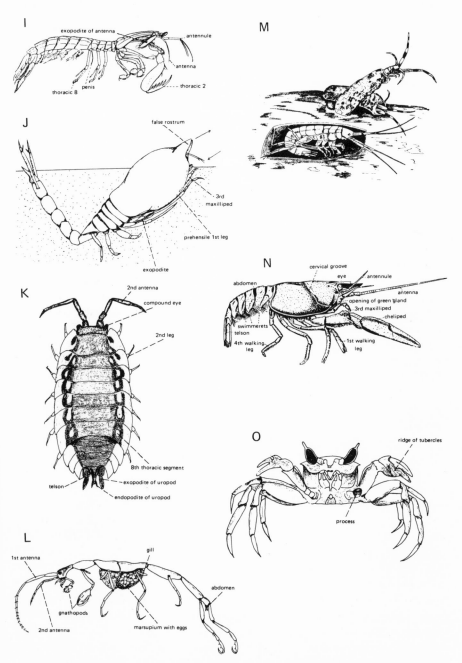

J, A cumacean, *Diastylis*. K, A terrestrial isopod, *Oniscus asellus*. L, A caprellid amphipod *Phthisica*. M, A tubicolous amphipod, *Cerapus*. N, A crayfish, *Astacus fluviatilis*. O, The ghost crab, *Ocypoda*. (A, B, C, E, F, J, K, L, M, and O are after various authors, from Barnes, 1968. D, G, I, and N are after various authors, from Borradaile *et al.*, 1958. H is after various authors, from Waterman and Chace, 1960.)

the home's "odor" with the owner's own, such explanations are labored, and some sort of knowledge of the terrain or other cues of the environment seems more likely. It has also been pointed out that despite apparently random wanderings during foraging, rock lobsters usually remain in fixed locales for many months at a time. This too suggests that they recognize the place as home. It has even been suggested that after major displacements lobsters can find their original homes miles away (Lindberg, 1955; Creaser and Travis, 1950; Bainbridge, 1961). The evidence for homing is not particularly compelling, but given our knowledge of such feats in turtles, birds, and perhaps some gastropod molluscs (Thorpe, 1956), it is perhaps not all that implausible.

Fiddler crabs dig holes in the sand just below the level of high tide and reside in these burrows during periods unfavorable for foraging, reproductive interactions, etc. Most of the time, the burrows are not private possessions, and when a crab leaves one to feed on the lower shore, he does not necessarily return to the same burrow. However, crabs of the species *Uca maracoani* go through behavioral phases in which certain sorts of activities are emphasized. In one of these, the so-called territorial phase, crabs take possession of a burrow for a day or so, ward off aggressors with threats and fighting if necessary, and when they feed some distance away at the water's edge, return in a straight-line path to the same burrow (Crane, 1958). Some sort of recognition behavior is clearly implied. The ghost crabs, genus *Ocypoda,* which retain burrows (Cowles, 1908; Schöne, 1961a), may wander a considerable distance laterally along the shore and yet are said to make a beeline for their burrow when frightened. This of course implies rather elaborate navigational abilities as well as memory. Daumer *et al.* (1963) have shown that *Ocypoda* can in fact learn to use polarized light cues to navigate between food and an artificial burrow.

Leaving aside the crustacean's ability to locate a precise home burrow or cave, rather specific geographical knowledge is also demonstrated in the ability of a number of littoral species to move directly toward or away from the sea (Schöne, 1961a; Pardi, 1960; Herrnkind, 1968). When such animals are taken from their home beach and placed in a circular pen with a view of the sky, they move in directions directly toward or away from where water ought to be. The sign of their movement depends on the motivating condition established by investigators. Where it has been investigated, both the sun and polarization of blue sky provide adequate navigational cues. In the case of the beach flea *Talitrus salter,* second-generation individuals born and raised in the laboratory and never exposed to shore or sun move in the direction of the native shore of their parents when tested for the first time (Pardi, 1960). This has adaptive value in these amphipods because the young grow up in the same general locale as the parents. This contrasts with the situation in fiddler crabs, which also display clear directional orientation on the shore (Herrnkind, 1968). Since the larval stages of the fiddlers are planktonic, and hence the young are presumably dispersed far and wide, inheritance is not a suitable means of supplying the requisite information. Of course, in principle it might be that those individuals with the wrong genetic background for the beach where they end up simply do not survive, but in the case of *Uca pugilator* it has been shown both in the field and in the laboratory using an artificial shore that direction is in fact learned (Herrnkind, 1968).

B. Social Behavior

A substantial number of studies (see, e.g., Hazlett, 1969; review in Lowe, 1956) have demonstrated the formation of dominance orders in laboratory populations of crustaceans.

Studies by Bovbjerg (1953) and Lowe (1956) on the crayfish are typical. Crayfish were placed together, four to a tank, and interactions were observed over many days. For those contacts between individuals that resulted in one of the two withdrawing, the quality of interaction (fight, strike, threat without fight, or lack of any clear-cut display) was scored, as was the outcome of the encounter (i.e., who withdrew). Figure 4 (after Lowe, 1956) illustrates the pattern typically observed for the interactions of the animal which ultimately became the dominant (α) individual of the group. The proportion of encounters from which it emerged victor increased over time (although high even initially), whereas the amount of overt aggression involved in the encounters declined markedly. It is clear in this example, as in many others, that some form of learning is involved, since time is required for the order

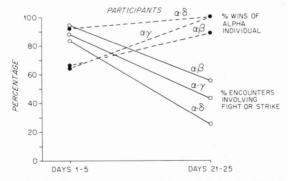

Fig. 4. Formation of a dominance hierarchy in the crayfish *Cambarellus shufeldtii*. The interactions of the dominant (α) individual with the second (β), third (γ), and fourth (δ) ranking individuals are characterized for the first and last 5 days of a 25-day interaction period. Note that α wins more but fights less as his superiority becomes recognized (data from Lowe, 1956).

to develop and stabilize and for the encounters to become ritualized. However, it is far from clear from such data just who is learning what.

The usual presumption is that animals learn by experience to recognize those specific individuals that they can and those that they cannot defeat in an overt fight. The capacity for such multiple individual recognition would of course imply the existence of rather well-developed perceptual and learning skills. In fact, however, all of the data available can be explained by assuming merely (1) that as a result of a variety of fights in a given situation, each individual learns about how often it can win and expresses this knowledge of personal worth in some sort of externally discriminable sign and (2) that other crayfish recognize such cues either innately or after some experience. There do not seem to be data available which would establish the thesis of individual recognition *per se*.

Experiments such as the above are carried out under abnormally crowded and artificial laboratory conditions. Nevertheless, since many of the species that have been studied probably do live at least for periods of time in fixed homes, it is likely that the formation of dominance orders does in fact prevent recurrent fighting by neighbors over shelter and over food when they meet while foraging. Indeed, conflicts over shelter have been observed in the species of crayfish studied by Bovbjerg (1953). However, the extent to which social learning actually plays a role in such encounters is not really known.

There are, however, a few observations on free-living animals which suggest that social learning does in fact have significant consequences. Crane (1958) reports that a fiddler crab which has been in an aggressive phase for an unusually long time may dominate other animals even after the phase has passed. Either the individual must be recognized as an individual or some subtle aspect of carriage or behavior must have become habitual. In either case, some sort of learning apparently is involved. Finally, it is known that male rock lobsters that are in sexual condition defend not only their own burrows but also adjacent ones from occupancy by other males. It seems likely that local males learn to avoid these territories.

C. Feeding Behavior and Food Preferences

Learning and memory are probably very important in the feeding behavior of many scavenging and predatory Crustacea. Patterns of search and hunting often necessarily depend on just what one is after. For example, a swimming crab which is after clams may systematically dig in a highly expert fashion; as it moves over a clam bed, it pauses every few inches and thrusts a claw straight down into the sand (MacGinitie and MacGinitie, 1949). If it gets a clam, it brings it up by running around its claw so as to keep the sand fluid. On the other hand, if the crab is after fish, it may lie quietly and alertly in ambush waiting for one to pass near. These are very different activities, and the persistent attention to either one for any period of time suggests that the animal at least can recall and maintain a chosen strategy.

Furthermore, the particular strategy adopted must presumably depend on the relative availability of various food types. Familiarity and abundance are the two crucial variables which may be expected to influence preferences, strategies, and so on. New foods are a risk in either the catching or the eating and must be dealt with cautiously. Scarce foods obviously demand a great expenditure of time and energy for their accumulation. Familiarity is a matter of learning, by definition, and the assessment of abundance will require memory whenever the animal is not literally surrounded by food.

That both of the above factors do influence the feeding behavior of Crustacea is shown by several observations on the rock lobster. Lobster fishermen report that the effectiveness of a given bait is dependent on the locality being fished; crushed mussels are good in areas with mussel beds but not elsewhere, and so on (Lindberg, 1955). D. P. Wilson reports that captive lobsters which normally would not eat hermit crabs will come to do so if no other food is available, and, once they do, the hermit crab becomes a preferred food (Wilson, 1949).

Learning of various sorts probably plays a role in the elaboration of feeding strategies as well as in the choice of strategies. It is not known whether the predatory activities of animals such as the swimming crab and mantis shrimps have in whole or in part to be learned, but it seems likely that at least the perfection of technique depends on experience (see discussion of development below). We have watched crayfish remove snails from their shells with a dispatch which we cannot help believing had something to do with prior experience, and we have also noted great differences in the ease with which individual crayfish capture and consume live brine shrimp (informal observations). Crabs sensing food juice in the water at first pass to their mouths any objects touching the claws. When the food reaches the mouth, it is rejected if it tastes bad (e.g., if it is filter paper soaked in alcohol). Eventually, however, the crab in such a situation learns (Luther, 1930) not to pass bad-tasting things to the mouth (a variety of appendages including the claws have chemoreceptors). Whether or not the basic reaction to food juices *per se* is also learned is not known; however, it is known that specific tastes can be associated with food. Thus the sensitivity of various crabs to cumarin has been increased as much as a thousandfold by training procedures (Luther, 1930; Wrede, 1929).

D. Sensorimotor Learning

As discussed in the introduction to this chapter, the feature of arthropods which was probably most important in leading to an advancement of their behavior over that of their polychaete ancestors and to a proliferation of

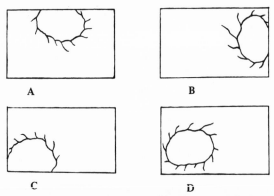

Fig. 5. Fabrication techniques in *Dromia*. Individuals were supplied rectangles of paper
in lieu of sponges (from Dembrowska, 1926).

learning abilities was the evolution of appendages which permitted precise
manipulation of objects. An interesting question, then, is the degree to which
the details of manipulative and other motor behaviors are themselves subject to learning. This is an especially interesting matter given our developing
knowledge of the neural circuitry of efferent control in decapods such as the
crayfish and lobster (e.g., Kennedy *et al.,* 1969; Davis, 1970; Larimer *et al.,*
1971). There has, however, been surprisingly little work on sensorimotor
learning in the Crustacea. Experiments with undoing various sorts of latches
to get into or out of closed containers would be easy enough to perform but
have not been done. That motor learning is possible, however, is suggested
by the following observations.

The sponge-carrying crab *Dromia* fashions for itself a covering made out
of sponge. Dembrowska (1926) found that if no sponge was available, the
crabs would make covers out of rectangular pieces of paper and that there
were several ways in which they proceeded. She categorized these ways into
four easily scorable types (Fig. 5) and tabulated the number of occasions on
which each of ten individuals proceeded a given way (Table II). The inter-

**Table II. Individual Differences in Cover Fabrication Technique
Among Sponge-Carrying Crabs,** *Dromia vulgaris* **(Data from
Dembrowska, 1926)**

| Fabrication method | Number of occasions on which each method was used |||||||||| |
| | Specimen No. |||||||||| |
	I	II	III	IV	V	VI	VII	VIII	IX	X
A	0	10	11	0	9	1	3	0	8	1
B	3	4	8	2	9	1	6	20	7	16
C	16	3	0	18	0	14	5	0	3	2
D	1	3	1	0	2	4	6	0	2	1

esting result was that each crab tends to have a personal way of constructing its case, and this suggests that learning may be involved. However, the possibility of genetic polymorphisms of case-constructing methods cannot be disregarded.

Dembrowska (1926) also showed that if sponges were buried under stones or snared on a hook, *Dromia* could gradually improve in its attempts to liberate them.

Several investigators have observed the effect of removing particular specialized appendages from crabs to see how they would manage with the remaining ones. Generally, there are reorganizations of behavior to compensate for such losses, but these are usually more or less immediate and therefore probably do not involve learning (Schöne, 1961a). However, on occasion some elements of learning may be involved, since the swimming of *Macropius* from which a swimming leg has been removed does show improvement with time (Kühl, 1933).

E. Approach and Avoidance Behavior

The real world is filled with good but not perfect correlations of events, and a measure of the adaptiveness of an animal's behavior is the extent to which it can capitalize on such contingencies in a way that facilitates survival. Large shadows often presage the approach of a predator, certain smells or tastes indicate the proximity of food, and so on. However, in some situations shadows may signify nothing of importance, and sources of food, although they are nearby, may be unavailable, poisonous, or in some other way unsatisfactory. Thus the ability to alter innate behavioral responses to some kinds of cues is very adaptive for most sorts of Crustacea, and such alterations are commonly observed.

For example, avoidance reactions to all sorts of normally frightening and flight-producing cues may habituate in animals held in captivity (see Table IV for references), and animals may become so tame that they can be fed by hand (Volz, 1938; Schöne, 1961a). On the other hand, animals caught in traps may become more wary than they were before (Appellöf, 1909; van der Heyde, 1920; Yerkes, 1902). Similarly, approach reactions, whether previously learned or innate, may habituate if they become inappropriate The ghost crab *Ocypoda* generally approaches and tries to catch objects which (apparently) look to it like flapping small fish washed up on shore; however, if an object is unsuitable for eating, for example, the shadow of a fish tied on a string out of reach, the crab will eventually learn not to try to catch it (Cowles, 1908).

Subtle cues, previously with little apparent effect on animals, can become strong determiners of behavior. As mentioned above, apparent sensitivity of a variety of crabs to cumarin can be greatly increased if the appearance of this substance accompanies the appearance of food (Luther, 1930; Wrede, 1929). Conversely, avoidance reactions to toxic substances may be increased by experience with them. When the amphipod *Gammarus* (Costa, 1966) is placed in a tube with a normal medium at one end and a toxic one at the other, the animal at first swims into the toxic region, apparently ignoring the low concentration of toxic substances at the boundary, but after a little experience it ceases to do so (Fig. 6). However, whether such changes are mediated by the nervous system is not clear. A final case of interest is the observation that *Ocypoda* becomes much more sensitive to tactile and vibratory stimuli when its eyes are painted over (Cowles, 1908). Learning is of course merely one of several interesting possible explanations of this phenomenon.

F. Role of Experience During Development

We know that in vertebrates the development of behavior is the product of a continual series of interactions between the developing animal as it is

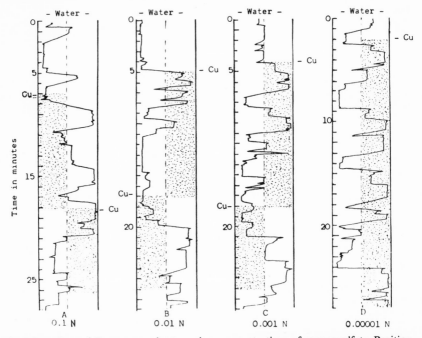

Fig. 6. Reactions of *Gammarus pulex* to various concentrations of copper sulfate. Position in the tube is shown horizontally and time vertically. Stippling indicates the toxic side of the tube (from Costa, 1966).

at a given moment and its environment. The widely known experiments of Held and Hein (1963) on the development of visually guided behavior in kittens are illustrations of the fact that experience can be critical to development in many ways that are more subtle than the wholesale learning of full patterns of behavior. We do not know why experience enters developmental sequences at the points it does. Whether it is a quite arbitrary accident of the particular history of an animal's evolution, or is of crucial adaptive significance for animals in a given niche, or is a device shaped by natural selection to get a complex organism out of an insufficiently informative genome is an open question.

However, whatever the reason, experience also plays a role in the development of crustacean behavior. This is seen particularly clearly in experiments by Hazlett on the shell-entering behavior of young hermit crabs. Adult hermit crabs which lack shells show specific reactions to appropriate gastropod shells; these include obvious interest, inspection, cleaning, and appropriate (and necessarily fairly precise) positionings, twistings, and so on, which get the animal into the shell. Adults raised in the laboratory may show almost all the behavioral units for getting into a shell without being able to utilize them in a successfully coordinated way (Hazlett and Provenzano, 1965; Hazlett, 1971). As in many of the cases studied in vertebrates, normal shell-entering behavior is not learned as such but does require that the animal have certain sorts of sensorimotor experience. This is illustrated by the following experiment (Hazlett, 1971). A number of freshly hatched zoeae larvae of a single brood were raised under several conditions which provided varying degrees of tactile and manipulative experience (see Table III). Group I simulated the normal sequence of developmental conditions, which includes a period in the plankton followed by life on the ocean floor. Group IV was at an opposite extreme. Those animals that were reared under group conditions (Table III) were frequently seen bumping into and grasping one another, while animals having sand available were

Table III. Role of Rearing Conditions on Development of Shell-Entering Behavior in Hermit Crabs (Data from Hazlett, 1971)

Group	Treatment from time of hatching till 1 week after molt to glaucothoe form (intermediate between larval and adult forms)	Treatment for next 3 days (all animals socially isolated)	Shell-entering errors of adults
I	Reared with numerous other specimens of same age	Piles of sand available	0.60
II	Reared with numerous other specimens of same age	No sand available	1.3
III	Socially isolated	Piles of sand available	2.0
IV	Socially isolated	No sand available	2.3

often seen to grasp and hold grains of the sand. It is clear from Table III that an early tactile-rich environment is important for efficient, relatively errorless shell-entering at a later stage of development. It is interesting that whereas shell-entering behavior is almost fully developed by the time the animal is ready for it, visual, as opposed to tactile, recognition of appropriate shell types appears to require actual experience with shells (Hazlett, 1971; Reese, 1963).

Isolation experiments (Hazlett and Provenzano, 1965) have shown that the normal highly ritualized social behavior of hermit crabs also depends on experience, and we may expect that where culturing methods permit, many more such cases will be found.

One of the relatively few cases where some ecological meaning can be given to the role of experience in development is in the orientation behavior of beach fleas and fiddler crabs discussed above. In the beach hoppers, which live out their lives near their places of birth, knowledge of the direction of the shore relative to solar cues is largely inborn, whereas in the fiddler crabs, which are dispersed by a planktonic zoeal stage, such knowledge is learned by each individual. However, even in the beach hoppers (Pardi, 1960) experience appears to play a role in sharpening orientation tendencies whose inborn component is relatively broad (Fig. 7). This is not surprising, for dispersal of the species obviously depends on the capacity to shift somewhat from an ancestral home.

G. Hidden Learning

Above have been discussed a variety of circumstances in which learning appears to be used by Crustacea in their daily lives. We have no trouble

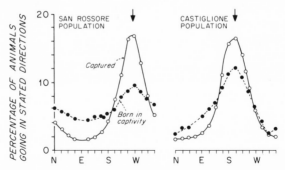

Fig. 7. Movements in response to desiccation in the littoral amphipod *Talitrus saltator*. Directions of movement in animals exposed only to the sky at sites away from the sea. Arrows at top show true direction of the home sea for two different populations of animals. Solid lines, freshly captured adults; dashed lines, specimens born in captivity (data from Pardi, 1960).

recognizing the ability to store the knowledge of the location of a home cave or the knowledge that a given kind of animal is good to eat as feats of learning, but we may easily fail to consider the great variety of rather ordinary things which animals do that in fact or in probability depend in a subtle way on some sort of learning phenomenon. It seems likely that the number of cases is legion; however, a small number of examples will illustrate the point.

1. The "Focusing of Attention"

Even the trivial but common event that an animal sets out to do something and perseveres at it through possible distractions demands a sort of "locking into a groove" which suggests that the animal in some sense "knows" and *remembers* what it is doing. A nice laboratory demonstration of this is the so-called telotactic behavior of various crustaceans in a two-light situation (Fraenkel and Gunn, 1961). This is shown in Fig. 8. In each case, the animal is photopositive and walks at all times toward one light or the other. *Both* lights are constantly in its field of view, and sometimes the animal appears to "change its mind." But at any one instant its attention seems to be focused on only *one* of the two "goals."

2. Orientation in Space

In animals that move around at will in an environment with fixed objects, it is important to remain oriented with respect to the static world. A swimming crab pursuing prey must immediately and accurately compensate for a shift of its own position due to a transient water current; a ghost crab heading for a flapping fish which it has sighted washed up on shore must efficiently return to its original path after detouring around a piece of debris

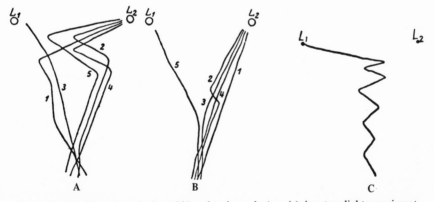

Fig. 8. Tracks of hermit crabs (a and b) and an isopod, *Aega* (c), in a two-light experiment. Each part of the track is directed toward one light only (from Fraenkel and Gunn, 1961).

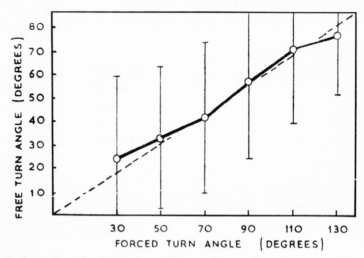

Fig. 9. Correcting behavior in *Armadillidium vulgare*. A forced turn in an alleyway is followed by a "free" turn upon emergence from the alley. The free turn opposes the forced one. Error lines are standard deviations (from Kupfermann, 1966).

on the shore; and any number of beach-living forms must continually know which way to run to go toward or away from the water. A variety of mechanisms are known which further such orientation, and almost all of them involve memory in some form.

For example, several sorts of insects and one crustacean type, the pill bug (Kupfermann, 1966; Watanabe and Iwata, 1956), have been shown to compensate for the sort of forced deviation from their path that might normally be caused by an obstacle (Fig. 9). This obviously demands that the animal remember either its original direction of travel or the angle through which it was forced to turn until such time as it can return to its original path. The duration of this sort of memory in Crustacea seems not to have been investigated, nor is it known whether the cues are proprioceptive or visual.

3. Optokinetic Memory

A similar phenomenon is well known in the visual domain. A great variety of animals, including various Crustacea, will do their best to maintain a constant orientation to their visual world. This is most dramatically demonstrated by the so-called optomotor or optokinetic reaction (see Pardi and Papi, 1961). To demonstrate this reaction, the animal is usually placed within a revolving, vertically striped drum which dominates its visual field. In such a situation it turns with the drum; if it is restrained and its eyes are mobile, they will deviate. The animal seems to do this not by fixating a particular contour and following it but by noting the drum's motion and then compensating for it. That memory is involved in this feat is perhaps not apparent until one realizes that for an animal to know that an object has moved, it must in essence remember where the object was a moment ago (Horridge, 1966a).

The crab *Carcinus* can detect the motion of a drum revolving at a rate of about one revolution in 3 days (Horridge, 1966b)! One way of executing such a feat of discrimination is to note the position of the drum at time 1 and then again at time 2 after it has had time to move a respectable amount. Such reasoning caused Horridge (1966a) to measure the persistence of "optokinetic" memory. The animal was exposed to a static drum for a period of time, the light extinguished, the drum moved during the dark period, and the drum then reilluminated. In order for the animal's eyes to deviate in the direction that the drum had moved, it had to remember the drum's original position. Figure 10 shows the extent of following as a function of time in the dark. There is clearly memory out to 10 min, and Horridge asserts that the total duration is at least 15–20 min. It might be supposed that the crab "remembers" by fixating some particular contour at a precise place on its retina and holding its eye fixed during the dark period much in the manner that many animals are said to remember by holding postural sets during delayed response tests. However, the eye in fact drifts a great deal during the dark period so that one must say that either the initial position of the drum or of the eye itself is remembered throughout the dark period.

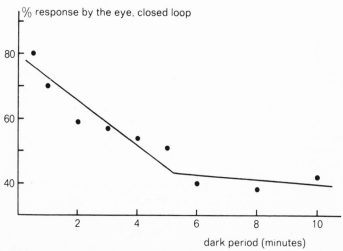

Fig. 10. Optokinetic memory in *Carcinus*. Eye movement as a percentage of drum movement after different periods of darkness during which the drum is moved (from Horridge, 1968).

It seems unlikely that such refinement of the optokinetic response is needed for maintaining moment-to-moment orientation in space. To what use then does the crab put its ability to perceive such slow movement? Ability (or willingness) to follow single lights is optimal in the range of 0.48–1.20 revolutions per day and falls off on either side of this (Horridge, 1966b). This suggests that perhaps the ability is used to perceive motions of the moon or sun (Horridge, 1966c). We have discussed above a number of cases of animals possessing sun compass reactions. It must be recognized that such animals know the correct heading relative to the position of the sun at *any* time of day. This means *either* that a veritable list of stimulus–response relations must be learned (i.e., where is the shore for each position of the sun—and time of day) or that the animal is capable of estimating and making use of knowledge of the angle which the sun's arc makes with the plane of horizon. In bees, it is known that the latter occurs and in fact that exposure to the sun only in the afternoon provides sufficient information for orientation at any time of day (Lindauer, 1961). Perhaps the ability of the crab to perceive the movements of the sun is involved in setting such a sun compass mechanism.

4. Perceptual and Motor Processes

Such analysis as has been done suggests that a great many routine perceptual and motor phenomena probably utilize short-term plastic properties of neurons in some degree. It is well known, for example, that neuromuscular junctions of Crustacea show a marked facilitation process in which the first of several closely spaced motor axoni mpulses primes the junction and permits subsequent impulses to release more transmitter substance than they would otherwise do (Wiersma, 1970; Dudel and Kuffler, 1961). As would be expected, command fibers which utilize such pathways formulate appropriately patterned trains of impulses in the relevant motor neurons (Wilson and Davis, 1965). Such temporal pattern codes are presumably techniques that animals with few motor neurons have evolved to gain precision control of muscle action. It was argued in a previous section that arthropods have probably made heavy use of temporal codes as opposed to spatial ones. In all such cases, decoding will require either integration over time or some more sophisticated form of memory.

Recordings from visually responsive interneurons of crustacean nervous systems suggest many examples of the importance of plasticity in perceptual analysis. Most obvious are "novelty" fibers, which respond to particular sorts of stimuli only briefly and then cannot be excited again until stimulation is discontinued for seconds or minutes. A typical example is fiber No. 30 in the optic nerve of the portunid crab *Podophthalamus* (Waterman *et al.*, 1964; Wiersma, 1970). This fiber responds for 3–4 sec to slow movement of an object in one small portion of the receptive field and then requires a 15-sec rest before responding again, although "naive" portions of the field remain responsive. This effect is probably not due to receptor adaptation, since other kinds of visually responsive interneurons retain undiminished responsiveness to stimulation of the same receptors.

A more fascinating and subtle example is seen in the behavior of "jittery movement" fibers of the crayfish. Extrapolating slightly from available experiments (Wiersma, 1970), such fibers seem to be able to respond to movements of a small object even while movements of the total patterned visual background relative to the animal do not elicit any response. In psychological terms, we would say that the animal distinguishes between figure and ground and does so even though the ground is moving relative to the animal's eye. This suggests that the animal's nervous system might recognize and remember the ground *as* ground. Experiments analogous to those of Horridge on optokinetic memory could tell whether this is so.

A much more speculative role for memory in perceptual phenomena derives from the notion that eyes having few ommatidia may in some cases be able to form fairly high-resolution central images by scanning objects and building up a spatial image over time. Gregory (1966) has given a fanciful account of the eye of the copepod *Copilia* in which a pair of lenses at the anterior corners of the carapace focuses an image on lenses and photoreceptors which are deep in the body and which continually scan back and forth across the image plane of the frontal lenses. Gregory suggests that the spatial pattern of light and dark is passed down the optic nerve as a time series much as in a television camera. A similar concept is hinted at by the Milnes (Milne and Milne, 1961), who report that several species of mantis shrimps inspect stationary objects at extremely close range by scanning them with the most sensitive but lowest-acuity portion of their highly mobile eyes. They appear as though they "were nearsighted and peering at details in order to comprehend the whole" (Milne and Milne, 1961, p. 424). It is clear that memory would be crucial to any image formation by scanning. The notion seems perhaps not so unlikely when we remember that (as Horridge, 1968, has pointed out) we ourselves can develop a very full impression of an object which we view moving behind a fixed, narrow slit. Scanning eyes coupled with appropriate memory and averaging circuitry could presumably serve as a way to utilize wide-aperture receptor elements to form high-resolution images under conditions of low illumination.

H. Phases and Rhythms

Crustacea, like other animals, show a variety of phases, states, or moods in their behavior. Stimuli which at one time or under one set of cir-

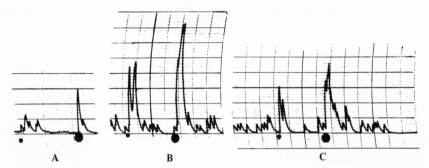

Fig. 11. The effect of "awakening" on a multimodal interneuron in the circumesophageal connective of the crayfish. Small dots, drop of water; large dots, jittery hand movement. A, Animal "asleep." B, Immediately afterward, animal alerted by the visual stimulus in A. C, Several seconds later without intervening stimulation (from Wiersma, 1970).

cumstances will evoke one pattern of response may under different circumstances have no effect or evoke much different behavior. Sometimes such behavioral states are under the control of known variables; sometimes they are not. But in all cases they are of interest to the student of behavioral plasticity.

Consider, for example, the daily patterns of activity of a fiddler crab such as the tropical species *Uca maracoani* studied by Crane (1958). During any low-tide period, most individuals tend to feed sequentially, engage in social interactions, repair their burrows, and feed some more, and finally retire underground. There are, however, wide variations in the amount of time devoted to the various activities on any given day. As Crane (1958) describes it, "many do not display at all, some remain inactive, others feed more than usual, and a few wander through the colony repeatedly dispossessing other males of their burrows and then passing on. Observations of the same individuals over successive days show that there is often an overnight shift in the dominant type of activity." In fact, six distinct phases can be identified, and while to some extent they are taken up in a fixed order, a principal characteristic is great variability of sequencing both among and between individuals. Phase switching probably does not involve learning as such; however, it is interesting to students of behavioral plasticity because it seems *a priori* so plausible that any sort of mechanism designed to alter behavioral state in a semistable fashion might well be utilized as part of the machinery of learning.

Other examples of change of state have known environmental triggers. A simple but not trivial example is the awakening of an animal from a state of "sleep" or repose into an excited or "awake" state. Figure 11, for example, shows the reaction of a multimodal sensory interneuron in the crayfish

Procambarus while the animal is "asleep" and after the animal has been "awakened" by a visual stimulus (Wiersma, 1970). The interest of such cases is not so much the switching of behavioral mode *per se* but the fact that often a new state (here, sleep or wakefulness) is *relatively persistent in time;* it is self-maintaining. Such persisting effects of transient stimuli are also especially marked in reactions to frightening stimuli. Thus feeding in hungry crayfish may be suppressed for many hours after a single fright (personal observation), and there are indications (Lindberg, 1955) that rock lobsters may evacuate their homes the night after encountering a diver.

The best-studied cases of change in behavioral state are those which occur periodically in phase with cycles of light or tide. General activity

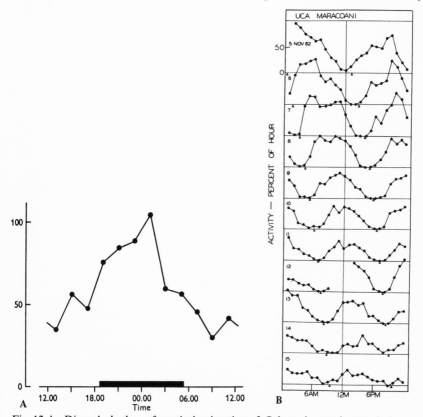

Fig. 12 A, Diurnal rhythm of vertical migration of *Calanus* in continuous darkness. Ordinate is the number of animals in the top half of the tank (average over 4 days). Black bar on abscissa indicates the nighttime, when plankton normally rise (from Harris, 1963). B, Tidal rhythm of activity in *Uca maraconi*. Average daily patterns of spontaneous motor activity in the laboratory. Triangles indicate the time of high tide at the beach where the animals were collected (from Barnwell, 1963).

levels, burrowing, foraging, and countless other behavioral manifestations are subject to such control. Indeed, almost the entire zooplankton, many of whose members are crustaceans, migrate to the surface at night and lie deep during the day. The striking thing about these cases where behavior is under the control of cyclic stimuli is that after some experience with a particular cycle the behaviors are performed at the proper time even if the animal is insulated from the relevant stimulus event. Figure 12 shows this for the vertical migration of *Calanus* (Harris, 1963) and the locomotor activity of *Uca* (Barnwell, 1963).

A psychologist might suppose that we have in these cases the operation of a conditioned response in which subjective time (either the time interval since the last stimulus or response, or the absolute time of a subjective clock) is the conditioned stimulus and light or tide is the unconditioned one.[2] However, the known facts are instead much more readily explained, as is well known, by the assumption that what is "learned" is the phasing of an internal cycle of almost invariant frequency; in other words, experience can lead to the resetting of an endogenous clock but can do little else.[3] The term *entrainment* rather than *learning* has generally been used to refer to this phenomenon. Entrainment, like classical and instrumental conditioning, results in a more or less permanent change in behavior and thus involves memory of a kind. Therefore, we may properly ask to what degree entrainment and learning might be fundamentally similar phenomena.

When a stimulus adequate to control the phase of an internal clock (e.g., dawn or light onset) occurs at a time different from that to which an organism has become accustomed, a change will occur in the phase of the internal cycle. The direction and magnitude of that change will depend on just where in the cycle the stimulus occurs and on its intensity. Without going into the details of the relationship between these variables, we may say that it is generally such as to promote fairly efficient restoration of the

---

[2]From such a supposition it would follow that protracted withholding of the light or tide stimulus would necessarily lead to extinction of the conditioned response. It would also follow that if time since the last response were the CS, then the time between successive responses could be arbitrarily chosen, whereas if the absolute time of an internal clock were the CS, responses could be conditioned to occur at more than one time during each revolution of the clock. Finally, if the CS-time were changed from X o'clock to Y o'clock, responses should gradually diminish at time X and increase at time Y; there would be no reason to expect a gradual progression of the time of response through various times between X and Y. In fact, all of these predictions are at variance with available data.

[3]It should be pointed out here that crustaceans offer a wealth of material for the study of cyclic behavior; indeed, much of the original work leading to our modern notions about internal clocks was carried out on members of this group (see review by Brown, 1961). Furthermore, such analyses have been carried to the neuronal level by Arechiga and Wiersma (1969), who have demonstrated circadian rhythms of excitability and activity in a variety of visually responsive interneurons of the crayfish.

normal phase relation between the timing stimulus (which the organism's timekeeping machinery always seems to treat as one instance of a periodically recurring stimulus) and the cyclic behavior of the animal (Pittendrigh, 1960). If, then, we wish to think of entrainment as a learning process, we may identify changes in the phase of the cycle under study as comparable to changes in strength or likelihood of a learnable response. Given this, we can proceed to ask whether the phase changes that occur during the process of entrainment have properties similar to those that are characteristic of learned changes in behavior. For example, is experimental extinction possible, does spontaneous recovery occur; is there forgetting (Woodworth and Schlosberg, 1956)?

Perhaps the most obvious point is that it is really rather difficult to see what the formal analogues of such characteristics might be in the case of entrainment. Nevertheless, there are, on the face of it, some rather suggestive phenomena. For example, if the fiddler crab *Uca pugnaxis* is placed under conditions of constant light, it appears to loose its diurnal rhythm of bodily color change (Brown, 1961). This could easily be thought of as analogous to forgetting or extinction (Woodworth and Schlosberg, 1956). A second example may be seen in members of the same species. If, after having been on a natural light cycle, the crabs are subjected to an artificial 12-hr day timed so that there is a dawn at true dawn and a dusk at dusk, they will advance the phase of their color change rhythm by one quarter of a cycle. If they are then placed in constant darkness at the start of a natural night, they at first maintain their altered phase setting but gradually revert to their original setting, which is in phase with true (but to them hidden) day and night (Brown, 1961). The student of learning might well consider this to be analogous to spontaneous recovery. Finally, it is not uncommon to find that when animals that have been exhibiting a cyclic response in uniform darkness are given a *single* stimulus out of phase with their previous cycle, they proceed to come gradually into phase with that stimulus through a sequence of phase shifts lasting many days (or cycles) (Harker, 1964; Pittendrigh, 1960); this phenomenon is not unlike what psychologists of learning refer to as reminiscence.

The previous paragraph seems to suggest that learning and entrainment might truly be related. However, while these various effects certainly have intriguing aspects, they do not in fact seem to be intrinsic characteristics of the entrainment process; rather, they seem to occur only in cases where behavior is controlled by multiple clocks that have somewhat different natural frequencies, are coupled to one another, and are unequally affected by particular external stimuli. Indeed, it seems to be true that phase changes in single clocks are intrinsically rather stable (Pittendrigh, 1960) and not subject to the various labilities that are characteristic of newly *learned* behavior. Although this statement is admittedly an inference rather than a fact, there does appear to be little basic formal similarity between learning and entrainment.

This is not to deny the possibility that certain cellular mechanisms are common to the two processes, but at present only speculation is possible. Nor is it to deny that entrainment of populations of endogenously rhythmic neurons might, by bringing units into greater synchrony, have profound effects on central excitability, on transmission through synaptic relays, etc., and thereby be involved in some of the forms of plasticity discussed elsewhere in this chapter. However, such notions are again purely speculative.

V. FORMAL LEARNING EXPERIMENTS

In this section will be reviewed what has been discovered about the learning abilities of Crustacea from formal learning experiments. Table IV constitutes a comprehensive list of such experiments. Our concern in the text will be with the nature of the capacities demonstrated.

Table IV. Annotated List of Formal Learning Experiments Classified by Subclass

Group	Authors	Animal	Characterization of study
1. Cephalocardia	None		
2. Branchiopoda	Blees (1919)	*Daphnia* (water flea)	*Instrumental learning:* photopositive animal learns to go toward darkness in order to escape from narrow tube; retention from one day to next
	Agar (1927)	*Daphnia* and *Simocephalus* (water fleas)	*Instrumental learning: failure* to learn position response for escape to water in 100–300 trials; errors punished by shock
	Applewhite and Morowitz (1966)	*Alonella* (water flea)	*Habituation* of shell-closing response; recovery occurs in minutes
3. Ostracoda	Applewhite and Morowitz (1966)	*Cyclocypris*	*Habituation* of shell-closing response; recovery occurs in minutes *Instrumental learning* of position response to escape light; forgetting within minutes *Classical conditioning:* light becomes CS for shellclosing; some controls for pseudoconditioning were run

Table IV (*Continued*)

Group	Authors	Animal	Characterization of study
4. Copepoda	Applewhite and Morowitz (1966)	*Paracyclops*	*Instrumental learning* of position response to escape light; forgetting within minutes
5. Cirripedia	Pieron (1910)	*Balanus* (acorn barnacle)	*Habituation* of withdrawal to shadows; see Gwilliam (1966) for preliminary physiological analysis of this effect
	Lagerspetz and Kivivuori (1970)	*Balanus*	*Habituation* of withdrawal to shadows; recovery requires more than 10 min
6. Malacostraca: Nebaliacea	None		
7. Malacostraca: Stomatopoda	None		
8. Malacostraca: Mysidacea	None		
9. Malacostraca: Cumacea	None		
10. Malacostraca: Tanaidacea	None		
11. Malacostraca: Isopoda	Bock (1942)	*Asellus* (freshwater isopod)	*Instrumental learning* of a position response to avoid punishment
			Instrumental discrimination of tactile cues to avoid punishment
		Porcellio (sow bug)	*Instrumental learning* of a position response to avoid punishment
	Watanabe and Iwata (1956)	*Armadillidium* (sow bug)	*Turn alternation:* animals choose to turn in a direction which opposes previously forced turns
	Thompson (1957)	*Armadillidium*	*Learning set: failure* to obtain improvement with repeated reversal of a position response
	Iwahara (1963)	*Armadillidium*	*Turn alternation*
	Sachs et al. (1965)	*Porcellio* (sow bug)	*Turn alternation*
	Hughes (1966)	*Porcellio*	*Turn alternation*
	Kupfermann (1966)	*Armadillidium*	*Turn alternation*
	Morrow (1966)	*Porcellio*	*Instrumental learning* of a position response; punishment compared to negative reinforcement without effect
	Harless (1967)	*Porcellio*	*Learning set: failure* to obtain improvement with repeated reversal of a position response

Table IV (*Continued*)

Group	Authors	Animal	Characterization of study
	Morrow and Smithson (1968)	*Porcellio*	*Instrumental learning* of a position response for water reward
12. Malacostraca: Amphipoda	None		
13. Malacostraca: Euphausiacea	None		
14. Malacostraca: Decapoda	Bethe (1898)	*Carcinus* (swimming crab)	*Instrumental learning: failure* to avoid punishment by suppressing innate (or previously learned) responses; few trials
	Yerkes (1902)	*Carcinus* (green crab)	*Instrumental learning* of paths through simple mazes for food or escape to water
	Yerkes and Huggins (1903)	*Cambarus* (crayfish)	*Instrumental learning* of position responses for escape to water
	Spaulding (1904)	*Eupagurus* (hermit crab)	*Instrumental learning* to go through doors into dark chamber for food
	Cowles (1908)	*Ocypoda* (ghost crab)	*Instrumental learning* of threechoice position response for escape *Classical conditioning:* dish which usually contains water becomes CS for gill-wetting response
	Drzewina (1908)	*Pachigrapsus* (amphibious crab)	*Instrumental learning* to go through door in glass wall to get to light
	Drzewina (1910)	*Clibanarius* (hermit crab)	*Habituation:* nude crabs cease attempting to enter sealed gastropod shells; unfamiliar shells re-excite response
	Schwartz and Safir (1915)	*Uca* (fiddler crab)	*Instrumental learning* of position response for return to burrow
	Van der Heyde (1920)	*Carcinus* (swimming crab)	*Instrumental learning* of path through complex maze
	Mikhailoff (1920, 1922)	*Pagurus* (hermit crab)	*Classical conditioning:* colored light becomes CS for withdrawal reflex, tactile stimulus to the head is UCS
	Mikhailoff (1923)	*Pagurus*	*Higher-order conditioning* of conditioned withdrawal reflex
		Leander (shrimp) and *Pagurus*	*Instrumental discrimination* of color; container of one color contains food
	Dembrowska (1926)	*Dromia* (sponge-carrying crab)	*Instrumental learning:* improvement at freeing sponge buried by rocks or snared by a hook

Table IV (*Continued*)

Group	Authors	Animal	Characterization of study
	Agar (1927)	*Parachaeraps* (crayfish)	*Instrumental learning* of a position response to escape to water, errors punished by electric shock
	Gilhousen (1927)	*Astacus* (crayfish)	*Instrumental learning* of a position response to escape to water
	Wrede (1929)	*Eupagurus* (hermit crab)	*Classical or instrumental learning* (?) of food-searching behavior in presence of a chemical substance previously associated with food; could be pseudoconditioning
	Luther (1930)	*Carcinus* (swimming crab) and *Eriocheir* (woolly-handed crab)	As above
	ten Cate-Kazejewa (1934)	*Eupagurus* (hermit crab)	*Classical conditioning:* CS is weak poke, UCS is strong poke; response is shell evacuation; could be pseudoconditioning
	Fink (1941)	*Pagurus* (hermit crab)	*Habituation* of fright reflex
	Datta *et al.* (1960)	*Gecarcinus* (Bermuda land crab)	*Learning set: failure* to obtain improvement with repeated reversals of a position response
	Schöne (1961*b*)	*Panulirus* (spiny lobster)	*Instrumental discrimination* of brightness cues; simultaneous presentation *Instrumental learning* of position response for escape to water
	Daumer *et al.* (1963)	*Ocypoda* (ghost crab)	*Instrumental discrimination* of polarization cues to navigate between food and home burrow
	Eisenstein and Mill (1965)	*Procambarus* (crayfish)	*Instrumental discrimination;* successive visual discrimination (go–no go) for food
	Hazlett (1966)	*Calcinus* (hermit crab)	*Temporal facilitation* of display
	Capretta and Rea (1967)	*Orconectes* (crayfish)	*Learning set: failure* to obtain improvement with repeated reversals of a position response
	Krasne and Woodsmall (1969)	*Procambarus* (crayfish)	*Habituation* of tail-flip escape response
	Chow and Leiman (in preparation)	*Cambarus* (crayfish)	*Instrumental discrimination:* photonegative animal must go from dark to light to avoid electric shock in a shuttle box

A. Changes in Level of Responsiveness

All animals exhibit a variety of changes of behavioral state for periods of time after certain sorts of stimulus or response events, and of course Crustacea are not exceptions to this rule.

In some cases, even single stimulus events may precipitate more or less persistent changes of state. We have mentioned some of these above. Some particularly clear examples are seen in certain interneurons of the crayfish (Taylor, 1970; Kennedy and Preston, 1963). Taylor, recording chronically from Wiersma's interneuron C4 in the circumesophageal connective, has shown that a variety of events can alter thresholds of this neuron to water-borne vibratory stimuli (Taylor, 1970). Many types of phasic visual stimuli readily cause a threshold elevation which typically lasts about 10 min (Fig. 13), while many other stimuli, or even apparently endogenous events, can precipitate threshold changes which may persist for hours.

Repeated presentation of a stimulus which evokes an "unlearned" response almost *invariably* results in some change in the likelihood of getting

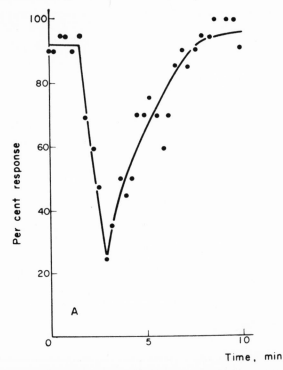

Fig. 13. Change in percent response of a vibration-sensitive neuron of the crayfish after a transient shadow stimulus (from Taylor, 1970).

the response. Changes in the decreasing direction (habituation) are most familiar, but increases ("facilitation" or "sensitization") are not uncommon.

An experiment of Halzett's (1966) on the aggressive displays of hermit crabs will illustrate the latter. The most common aggressive displays of *Calcinus tibicen,* the species used, are upward and outward movements of either the claws ("cheliped extension") or the legs ("ambulatory raise"). Hazlett presented animals with models in one of these postures some 200 times over several days and then, 3 hr after the last presentation, put the animals in groups to see whether the display an individual would utilize was affected by the type of model to which it had been exposed. There was a clear tendency during "training" for animals to "answer" a given posture of the model in like kind, and they continued to use the same posture 3 hr later during the test (Table V).

The phenomenon of habituation has been studied particularly intensively. It is almost invariably seen in frightening and arousal-producing situations at all phyletic levels.

Fink's (1941) experiment on hermit crabs is typical. Each day, animals were placed individually in a test dish. When the dish was struck sharply, a crab at first withdrew into its shell. After it reemerged, the dish was struck again and so on until three successive failures to withdraw had occurred. Each of 30 animals was run in this way for 17 days. Figure 14 shows that mean trials to criterion went down markedly over days. Many animals simply ceased responding altogether during the last days of the experiment.

Such habituation is a highly ubiquitous phenomenon; however, the speed with which it develops and, more interestingly, the extent to which it persists over days vary widely across species and situations. Thus it is clear that in the present example the habituation persists in considerable measure at least from one day to the next, whereas in acute experiments on the habituation of giant fiber escape reactions of the crayfish (see below) recovery from habituation requires only minutes. Whether the memory phenomena

Table V. Temporal Social Facilitation in the Hermit Crab *Calcinus tibicen* **(Data from Hazlett, 1966)**

	Percent responses of same type as model	
Model presented	During presentation[a]	3 hr later during interactions
Cheliped extension (40 specimens tested)	76.7	65.5
Ambulatory raise (40 specimens tested)	86.3	60.6

[a]In each type of test, the effect of model presented was significant at $p < 0.01$ by a 2 × 2 χ^2 contingency test.

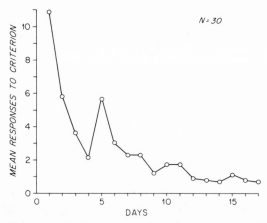

Fig. 14. Habituation of the withdrawal reflex in *Pagurus longicarpus* (data from Fink, 1941).

involved in examples at these two extremes are basically similar should be questioned.

Usually, a specific reaction is observed in habituation experiments, and it is tacitly assumed that the failing reflex is somehow selectively reduced in strength. In fact, however, it is often the case that a whole cluster of behavioral changes occurs. A variety of animals (Doflein, 1910; Volz, 1938; Schöne, 1961a) not only cease to escape from the experimenter but may also become willing to accept food and may become very tame or even aggressive (personal observation on crayfish). It seems likely that in such cases it is "fear" which is habituating and not a specific reaction, and neurophysiologists who seek changes in the specific reflex pathway of a particular flight reflex are liable to be disappointed.

Almost always, in Crustacea as in other animals, habituation of a response is at least in some measure specific to the stimulus evoking it or the situation in which it occurs. Fink reported that in the above experiment the crabs had little tendency to withdraw to vibrations when they were in the home dishes where they were normally fed; presumably, the flight reflex had already habituated in that context. Doflein (1910) described the taming of the shrimp *Palaemon* as being reversed by a change of situation. Drzewina (1910) and later Hertz (1932) showed that naked hermit crabs would eventually give up trying to enter plugged snail shells of one kind but would proceed again when a new and also plugged type of shell was introduced. Crayfish whose tail-flip escape response has habituated to tactile stimulation of one abdominal segment may still have a considerable tendency to respond if a different segment is stimulated (Krasne and Woodsmall, 1969). All of these examples show quite clearly that the animal not only learns to cease responding but also learns something about the eliciting stimulus. Of course, this is adaptive. It also shows that in a sense there is an element of association between the stimulus and response even in habituation learning. Therefore, it is not at all obvious that the character of such learning or its physiological basis will necessarily be different from that of conventional associative learning.

B. Associative Learning

1. Classical Conditioning

The classical Pavlovian paradigm of associative learning has very seldom been utilized with Crustacea (Table IV). We shall consider the defining characteristic of classical conditioning to be an approximate isomorphism between the conditioned and unconditioned responses. One of the few formally unambiguous examples comes, quite properly, from a series of papers by a Russian worker, Mikhailoff (1920, 1922). Naive hermit crabs of the species *Pagurus striatus* draw rapidly into their shells when they are touched between the eyes. Mikhailoff's demonstration consisted of consistently turning on a colored light at the moment of tapping an animal. The animals never responded to colored lights by withdrawal before such pairing, but they came to do so quite reliably as the pairings were presented over a number of days (Fig. 15). Training was carried out with either red, red-orange, green, or yellow lights, depending on the subject. From time to time throughout training, the animals were tested with colors to which they had

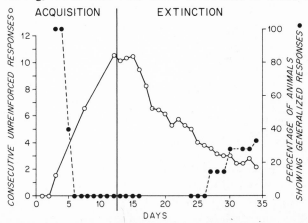

Fig. 15. Conditioning of the withdrawal reflex in *Pagurus striatus*. During acquisition, an average of 30 pairings of a colored light and tactile stimulation was given per day. Responses to the light alone were tested intermittently, and when one occurred further such trials were run at 30-sec intervals to determine the stability of the conditioned reflex (left ordinate, open circles). From time to time, a color slightly different from the CS was tested (right ordinate, filled circles) (data from Mikhailoff, 1922).

not been trained, and whereas colors similar to the training stimulus pro-
voked reactions in such cases early in training, they did not do so later on.
When reinforcement was ceased, extinction gradually occurred (Fig. 15).
Explanations of these data in terms of sensitization phenomena rather than
association *per se* are possible but quite tortuous and probably not correct
in view of the totality of accomplishments of these animals.

2. Instrumental Learning

The great bulk of associative learning studies with Crustacea have been
on the formation of simple position habits in two-choice situations such as
T- or Y-mazes (see Table IV). The original experiments of Yerkes, one of
which was described in an earlier section, are typical. Motivation for learning
has been escape from heat or dryness, return to home containers, and occasionally food; in some studies, electric shock has also been given as punishment for an erroneous choice. In some instances, cues such as brightness or
texture have been added to those of position, but this has never actually been
demonstrated to facilitate learning. While there may be some members of the
class that cannot learn in this sort of simple situation (e.g., see Agar, 1927),
the burden of proof at this point is probably on the purveyor of a negative
result. On the other hand, it must be admitted that the great bulk of the
work that has been done has been on malacostracans (Table IV). Furthermore, it should be pointed out that a great many, though by no means all, of
the studies available are subject to the criticism that they appear not to have
taken adequate account of the possibility that individuals may be able to
"learn" by laying and following appropriate sorts of chemical trails. Explanations in such terms sometimes need to be rather involved, but they
probably should not be rejected out of hand given what is known of insect
pheromones.

Nonpositional discrimination studies are much rarer (Table IV) but
provide the most definitive evidence of associative learning. Schöne's (1961b)
study of a simultaneous brightness discrimination in the rock lobster, *Panulirus argus,* is a good example. The animals were rewarded by return to water
for going toward the correct of two plaques differing in brightness of illumination. The position of the correct plaque was varied from trial to trial
so that the animals had to respond to the brightness cue *per se.* Their performance curves are shown in Fig. 16. There is no doubt that the animals
learned although their rate of acquisition was rather slow.

Particularly impressive is an early experiment of Mikhailoff's (1923).
Four shrimp of the genus *Leander,* and four hermit crabs, *Pagurus striatus,*
were presented ten times a day with a pair of cylinders, one red and one green,
each with a hook inside. On alternate trials, the hook within the cylinder of
one color (always green or always red for a given animal) was baited with

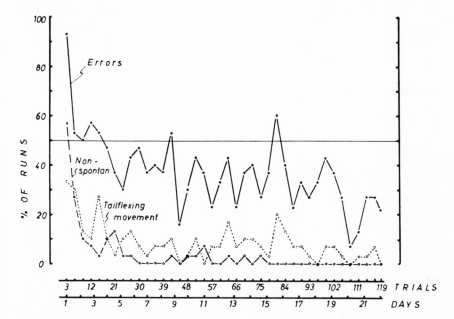

Fig. 16. Brightness discrimination learning in *Panulirus argus*. Erroneous runs are those in which the animal went initially toward the incorrect stimulus. Nonspontaneous runs are those in which the animal remained initially near the starting point. The exit to the maze was beneath the brighter window (lobsters naturally prefer the darker one). Ten animals were used (from Schöne, 1961*b*).

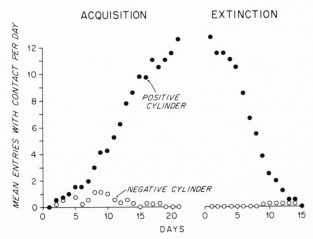

Fig. 17. Color discrimination learning in *Pagurus* and *Leander*. Graphed are responses on test trials where no food was present. Eight animals were used (data from Mikhailoff, 1923).

food, which the subjects easily found and devoured. The question of interest was the behavior of the animals on trials where no food was available. Figure 17 shows that they readily learned to enter a cylinder similar to the one in which they had been fed and to search for food on the hook within, even though the actual cylinders and hooks used for these tests had never been touched by food.

C. Extent of Attainments

We have seen above that many members of the class Crustacea are certainly able to learn. Here we wish to assay how well developed that ability is—how rapid is acquisition of a habit, how precise and errorless is the ultimate performance, how well is what is learned remembered, how much information can be stored in a given situation, and so on. In making such evaluations, it must, of course, be understood at the outset that in laboratory learning studies the manner of dealing with the animals and of performing the experiments is always, through our ignorance, suboptimal. However, if we bear this limitation in mind, the results are certainly of some interest, at least in giving the lower limits of an animal's capacities.

In Table VI are summarized certain salient features of a number of available studies.

1. Memory

Addressing first the question of memory ability, it is clear from the few cases where memory has been evaluated that simple spatial discriminations can be retained for at least 10–15 days (Table VI). Even in cases where memory has not been explicitly evaluated, it is clear that what is learned can be recalled for more than 24 hr, since performance generally improves over successive days of testing. The most impressive feat of memory shown is that of the hermit crab *Pagurus striatus,* whose conditioned withdrawal response to a colored light persisted despite daily extinction trials over a period of several months (Fig. 15 and Mikhailoff, 1922). The retention abilities of those Crustacea that have been tested thus appear to be comparable to what one sees in experiments on mammals.

However, it should be noted that this conclusion really pertains only to the Malacostraca. There are indications in the few experiments that have been done on representatives of other groups (Table IV and Applewhite and Morowitz, 1966) that memory might be less well developed elsewhere in the class.

2. Precision

When one considers the precision of learning (or at least of the *perfor-*

mance of learned behavior), Crustacea seem not to fare so well (Table VI). Ten to twenty percent errors are typical of two-choice situations where animals are trained for several hundreds of trials distributed over many days. Furthermore, investigators who wish for the purposes of their experiments to run animals to some criterion of attainment generally use rather weak criteria such as eight out of ten correct rather than demanding long runs of consecutively correct responses, and it must be presumed that this strategy has its reasons. Of course, even very imperfect learning may still be of great selective value.

3. Storage Capacity

Attempts to assess the information-storage capacity of the human brain and to compare it to the numbers of neurons or synapses it contains are common and instructive, if not fully convincing. Similar analyses for invertebrates with their much lower cell counts would be interesting, but it is even more difficult to envisage meaningful ways of proceeding with the analysis of storage capacity than it is in the human case. Nevertheless, one could in principle ask questions such as how many discriminations can be held simultaneously

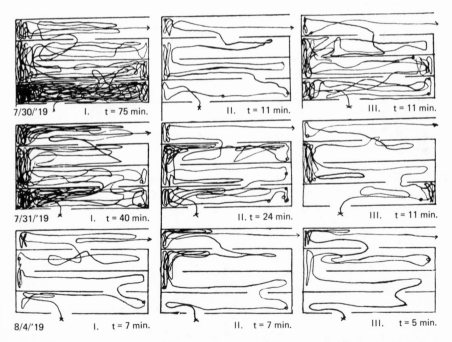

Fig. 18. Complex maze learning in *Carcinus maenas*. The three rows show the first 3 days of training of one specimen; the three trials of each day run from left to right. The general improvement is evident (from van der Heyde, 1920).

Table VI. Quantitative Data on Selected Learning Studies

Study	Animal	Type of situation	Terminal performance	After how many trials?	At how many per day?	Is performance asymptotic?	Recall
Yerkes (1902)	Crab	Maze slightly more complex than T-maze	Imperfect	50	4	?	?
Yerkes and Huggins (1903) Expt I	Crayfish	Two-choice position	90% correct	60	2	No	73% correct after 14 days
Yerkes and Huggins (1903) Expt II	Crayfish	Two-choice position	88% correct (not terminal) / 98% correct	150 / 250	10	No / ?	?
Schwartz and Safir (1915)	Fiddler crab	Two-choice position	80% correct	200	20	No	80% correct after 10 days
Mikhailoff (1922)	Hermit crab	Classical conditioning of withdrawal response to color CS	Capable of 10 consecutive unreinforced responses; no responses to colors other than CS	Approx. 300 (?) pairings	Approx. 30 (?) pairings	?	After 22 days extinction, capable of 2.2 unreinforced responses following 24 hr spont. recovery; responses to colors other than CS start to appear after 14 days of extinction
Mikhailoff (1923)	Hermit crab and prawn	Simultaneous red–green discrimination	See Fig. 17				Discrimination persists in part through 10 days of extinction

Gilhousen (1927)	Crayfish	Two-choice position	88% correct	60	5	Yes	?
Datta et al. (1960)	Bermuda land crab	Two-choice position	80% correct	360	10	Yes	?
Schöne (1961b)	Rock lobster	Two-choice position	92% correct	27	3	No	?
Schöne (1961b)	Rock lobster	Simultaneous brightness discrimination	80% correct	120	5	No	?
Morrow (1966)	Sow bug	Two-choice position	80% correct on 2 successive days (criterion)	89	10	?	?
Harless (1967)	Sow bug	Two-choice position	80% correct (criterion)	46	10	?	?

and whether there is a point where confusions become rampant; however, such questions have not been investigated. The closest thing available, and it is not very close, is studies on multiple-choice-point mazes. There have been only two such studies, and in neither did the animals ever run through the mazes in the expeditious manner often seen in rat studies. Even when few errors are made, long pauses are fairly common, and investigators usually choose to prod the beasts to action when these pauses become too protracted. Nevertheless, there *is* clear improvement, as is illustrated by the behavior of the swimming crab, *Carcinus maenas,* shown in Fig. 18 (van der Heyde, 1920). The average time to the exit for the eight animals run in this study decreased from 60 min to 3 min over 15 trials at three per day, which in fact seems rather impressively fast learning. Unfortunately, however, van der Heyde does not present quantitative error data, and it is this that is required to determine how good the animals are at storing information. Even error data are difficult to interpret in most mazes, because the learning of simple strategies for the efficient use of recent memory (i.e., good search strategies) can *greatly* improve performance in a complex maze. For this reason and because animals which make errors and explore blind alleys may in fact have to learn a great deal more about a maze than simply each of the choices on the *correct* path, it is very difficult to determine information-storage capacity from complex maze learning ability.

An unpublished study done in the author's laboratory by A. Barnebey (the only complex learning study available except van der Heyde's) may be taken to illustrate some of the problems as well as to indicate in some slight measure the animals' degree of attainment. Barnebey trained his animals for three to five trials per day first on a fairly simple maze (Fig. 19, maze *A*) and then on a more complex one (Fig. 19, maze *B* or *C*) to which he hoped there would be some transfer of training. His method of scoring errors was to divide the mazes into a number of zones shown by the dashed lines in Fig. 19 and to count the number of times the animal moved into a zone which would take it *away* from the goal. The errors for typical paths through the maze are shown by the arrow heads in the figure. The error scores for all subjects are shown in Fig. 20. There is a clear drop in errors on the first, simple maze (Fig. 20A), but before perfect performance was obtained the animals were transferred to the second maze (Fig. 20B). There does appear to have been some positive transfer to this second maze, but performance then deteriorated before it again improved. Apparently, the animals had some unlearning to do as well. It is difficult to say how well the animals were really doing at information storage in terms of the task set for them, because it is hard to know what behavior an adaptive search pattern would yield.

In maze *B*, with the left door open, for example, random wandering throughout the maze would lead to a very large number of expected errors

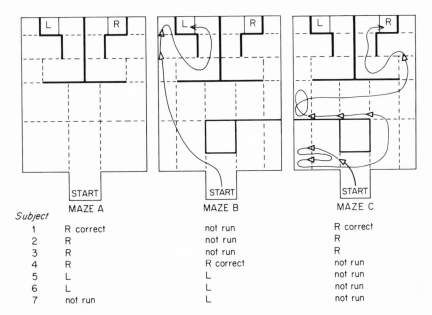

Subject	MAZE A	MAZE B	MAZE C
1	R correct	not run	R correct
2	R	not run	R
3	R	not run	R
4	R	R correct	not run
5	L	L	not run
6	L	L	not run
7	not run	L	not run

Fig. 19. Mazes used by Barnebey (see text). Below the mazes, the successive treatments given each subject in the above mazes are shown. In mazes B and C are shown paths with errors marked by triangles to illustrate the convention used for scoring errors.

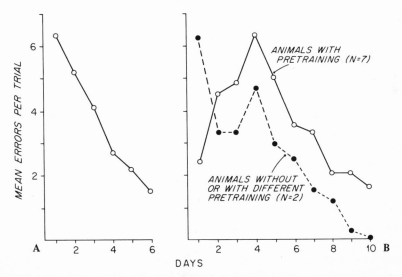

Fig. 20. Performance in the mazes of Fig. 19. A, Performance in simpler maze A (six animals). B (solid lines), Performance of the same animals in the more complex mazes B or C. B (dashed lines), Performance of animals 4 and 7 on maze B (unpublished data).

even under the assumption that the animal does not return to a *cul-de-sac* where he has just been. On the other hand, an animal learning *merely* to keep his left side in contact with a wall could get to the goal with only two errors (see Fig. 19, dashed lines of maze *B*), which is about the number of errors that Barnebey's animals made on the average at the end of their training. Of course, the animals obviously had not yet reached an asymptote of performance. Two of Barnebey's animals are of special interest; No. 7 did not have to unlearn anything, and No. 4 had a different pattern of prior learning relative to his ultimate objective (see Fig. 19). Figure 20B, which plots the performance of these animals on maze *B* as a dashed line, suggests that crayfish are in fact capable of perfect performance in this sort of maze after not all that much training.

The moral of this story, however, is that if one wishes to estimate information storage by maze-learning experiments, then one should utilize mazes which present a series of symmetrical choices in fixed order, and performance should be scored in terms of behavior at those choice points. Useful data on information storage capacity could also be obtained from experiments on visual pattern discriminations like those carried out by Herz (Herz, 1933, 1934; Carthy, 1958) on bees or from experiments on multiple olfactory discriminations.

4. Generality of Learning Capacity

In the introduction to this chapter, we raised the question of whether the ability to learn is a capacity conferred on an organism as a whole or on particular domains of an organism's behavior. There are indications that, contrary to the belief of early Pavlovians and behaviorists, even mammals are not perfectly general learning machines. Garcia *et al.,* (1968) have shown that rats can learn that particular sights and sounds presage pain but they cannot easily learn of a relationship between such stimuli and impending sickness, whereas tastes and smells are useful warnings of illness but sights and sounds are not.

Whatever the situation for vertebrates, there are indications that the learning capacities of many invertebrates, including the Crustacea, are to some degree behavior specific. Bock (1942) found that whereas the terrestrial isopod *Porcellio* could learn a position habit in either a horizontal or "vertical" T-maze, the freshwater isopod *Asellus* could only learn in a horizontal maze. Harless (Marion Harless, personal communication) has speculated that *Porcellio*'s ability in the vertical maze might be related to its habit (Brereton, 1957) of climbing trees. Be this as it may, it suggests that learning in crustaceans may be limited to special domains. We would also suspect that Agar's (1927) inability to teach *Daphnia* or *Simocephalus* position responses in a T-maze (see Table IV) may well be a reflection of the uselessness of position response learning for planktonic animals.

A seemingly remarkable compartmentalization of learning ability is seen in the approach and avoidance behavior of crayfish and lobster. The animals seem quite expert at avoiding an adversary by swimming away from him, which they do by rhythmic flapping of their abdomens, and their visual–motor control is good enough that they expertly swim around obstacles in their path (Lindberg, 1955; personal observation). However, they normally seem quite unable or unwilling to utilize this behavior to catch food which is out of their reach (Lindberg, 1955; personal observation). We have dangled pieces of liver just out of reach of very hungry crayfish, and though they consistently try to scurry up the glass sides of their aquaria to get the food, they seem quite oblivious to the possibility of swimming to it. In a few instances, we have seen *extremely* excited individuals start to swim under these circumstances, but they do not aim for the food, and although they may eventually run into it once they are off the bottom, it seems to be only by accident. Although swimming toward food might eventually be learned, this certainly does not happen easily.

From the point of view of *adaptation,* behavior-specific learning capacities are not surprising in animals with highly specialized modes of existence; however, their occurrence in *these* cases raises the general question of whether all learning abilities are specifically and separately evolved adaptations. Properly conceived comparative research should be able to answer this fundamental question.

D. "Higher Mental Processes"

Having seen that many crustaceans are well able to learn, we may inquire whether they might also display some of the more advanced sorts of learning phenomena which, at least in the view of some psychologists, play a role in mammalian intelligence.

1. Secondary Reinforcement

The phenomenon of secondary reinforcement (token reward, or higher-order conditioning) is crucial to the liberation of behavior from primary reward and is therefore essential to most of the complex learned behaviors of intelligent animals. To the author's knowledge, the only attempt to investigate whether Crustacea can learn by secondary reinforcement was an experiment on three specimens of the hermit crab *Pagurus striatus* by Mikhailoff (1923). The general methods of training were described above in the section on classical conditioning. In the present experiment, the CS (a colored light) for a well-established CR (withdrawal of the animal into its shell) was paired with a new CS (a light of a different color). From time to time, the original CS was reinforced by the unconditioned stimulus (a tap between the eyes). Table VII shows the number of pairings to the first conditioned re-

Table VII. Higher-Order Conditioning in *Pagurus*, with Data on Normal
Conditioning Given for Comparison (Data from Mikhailoff, 1923)

Animal	Primary CS.	Approximate number of pairings before first CR[a]	Higher-order CS	Number of pairings before first CR
1	Red	45	Green	154
2	Red-orange	51	Green	173
3	Green	135	Red	197

[a]Data in this column, which are given for comparison only, are means for CSs of the same color in an earlier experiment. Mikhailoff does not give initial learning data for the present experiment.

sponse for both the primary and the higher-order conditioned stimuli. The
N of this experiment is small, but the results seem reasonably convincing—
although there are certain controls that one might wish Mikhailoff had run.

2. Partial Reward

Although not normally considered a sign of advanced intellect, the
phenomenon of high-level performance during partial reinforcement and
of unusually slow extinction following it seems intuitively, at least to the
author, as an indication of some considerable mental attainment, for it
demands that the animal learn to ignore events which would normally
discourage it. Perhaps in holding this opinion the author is merely reacting
to the difficulty which learning theorists have had explaining the phenomenon. Indeed, experiments on annelids suggest the this opinion is
wrong. Oddly enough, however, we know of no test of the effect of partial
reinforcement on a crustacean.

One of the more imaginative explanations of slow extinction after partial reward in mammals is based on the idea that cessation of responding
during experimental extinction is an effort on the part of the animal to avoid
the unpleasant aspects of the frustration which results when an "expected"
reward is not forthcoming (Amsel, 1958). Given this, the argument is made
that animals on schedules of partial reinforcement learn to continue responding in the face of frustration, which in fact becomes a discriminative stimulus
for reward. We raise this entertaining notion because there are some data
suggesting the possibility of frustration effects even in the lobster. The data,
although not the interpretation, are Schöne's (1961b). If a lobster in a
solvable visual discrimination situation makes an incorrect choice, he will in
21 % of the cases express what we may interpret as his displeasure by executing a tail-flip escape response. Furthermore, if he is in an unsolvable
situation, so that, in effect, his behavior is on a 50 % schedule of partial
reinforcement, and hence frustration is presumably more extreme, he will

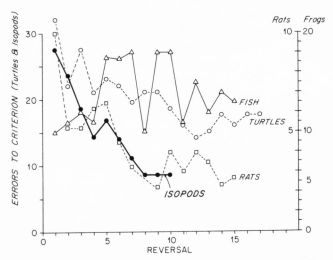

Fig. 21. Repeated reversal learning in *Porcellio scaber*. Errors to a criterion of nine out of ten correct responses are plotted for successive reversals of a position habit (heavy solid line) (after Morrow and Smithson, 1969). For comparison, reversal data from several vertebrate types are shown (after Bitterman, 1965).

execute escape responses on an even greater proportion $(32-43\%)[4]$ of those trials on which he makes an incorrect choice. Perhaps, then, even invertebrates are not without emotion, though whether this speculation belongs in a section on higher mental processes could be debated.

3. Learning to Learn

Ability to learn and form discriminations does not provide a good index of either evolutionary status or intelligence, at least as we intuitively estimate it, among the mammalian orders (Warren, 1965). A simple maze or discrimination is learned essentially as easily by a rat as by a man. The capacity to form learning sets, however, has been found to correlate fairly well with other indices of intelligence (Warren, 1965; Hodos, 1970). One of the simplest ways to test animals for the ability to form learning sets is to examine their performance during repeated reversals of a single discrimination problem. Animals that are repeatedly subjected to reversals in a spatial or visual discrimination task should optimally come to reverse their behavior as soon as a single nonreward is encountered; they should then hold to the new correct response until the first trial where it does not pay off; and so on. Typically, primates can learn to behave in this way (Warren, 1965). Less intelligent mammals such as the rat never do quite this well but do come to

[4]Comparison of either of these proportions to 21% is significant at $p < 0.001$ using trials on which errors were made as the population for a χ^2 test.

make only a few errors per reversal (Fig. 21). It should be noted that reversals are usually much more difficult than original learning (not shown in Fig. 21); this is because a previous habit must be extinguished as well as a new habit learned. In one of his pioneering papers on learning in Crustacea, Yerkes claimed that crayfish seemed to show improvement in successive reversals. However, his data on this point were far from convincing, and a number of subsequent attempts to obtain this effect in Crustacea have been unsuccessful (see Table IV).

Recently, however, Morrow and Smithson (1969) reported that they found improvement during repeated reversals in the isopod *Porcellio scaber*. The animals were trained in a T-maze, initially away from the side of their pretraining preference, to a criterion of nine out of ten correct responses with ten trials at 20-min intervals being given each day. The results are shown in Fig. 21 alongside the data for several vertebrate types. When compared with the failure of fish to show improvement, Morrow and Smithson's data may be regarded as rather odd. An important control in all such experiments is an examination of the reversal behavior of animals that have had maze experience but not experience in reversals *per se*, for increases in tameness and a variety of other nonspecific effects could give apparent improvement in reversal learning which we would not wish to attribute to the "intelligence" of the subjects (Bitterman, 1965). In the absence of such a control, Morrow and Smithson's pretty data are unfortunately difficult to interpret.

VI. PHYSIOLOGICAL ANALYSIS OF LEARNING

One of the great allures of investigating plastic processes in Crustacea and other arthropods is that we know that these animals learn, we know that for a variety of reasons they are very adaptable to neurophysiological investigations, and we know that, although they are very complicated, their behavioral neurophysiology is probably going to be much easier to make sense out of than that of mammals. Thus we have faith that through diligence, persistence, and intelligent investigation the arthropods or perhaps other invertebrate groups may yield up some of the great secrets of nervous plasticity. In this section, we shall explore some of the steps that have been taken in this direction and a few ideas which might bear some fruitful insights. However, it is this author's opinion that *as yet* no one has really come close either in the Crustacea or in any other invertebrate group to finding a way of investigating the sorts of learning phenomena that really interest us most.

There are essentially three broad approaches to the physiological study of learning and memory, and we shall consider them each, as they bear on the Crustacea, in separate sections.

A. Method I: Raw Mechanisms and Analogues

During physiological investigations of nervous systems, one frequently comes upon phenomena which one recognizes could be used in the design of machinery for learning and memory. Of course, the fact that a particular phenomenon exists does not necessarily mean that evolution has fashioned it into a device for learning or retention; the phenomenon's actual function is always a question for separate investigation. However, the fact remains that such phenomena are eminently worthy of study by students of behavioral plasticity.

1. Facilitation and Antifacilitation

Two phenomena which have particularly attracted the attention of people interested in learning are the changes in synaptic efficacy known as facilitation and antifacilitation. Figure 22A (from Bruner and Kennedy, 1970) illustrates the phenomenon of antifacilitation at a crustacean neuromuscular synapse. When a motor giant axon to the phasic flexor muscles of the crayfish tail is stimulated repeatedly about once per minute, a marked diminution of excitatory junctional potentials occurs. Recovery, which is fairly slow in neurophysiological terms, requires some tens of minutes for completion. The functionally opposite phenomenon, which is called facilita-

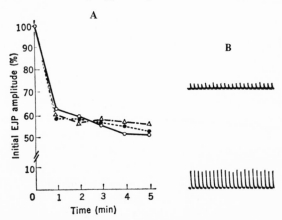

Fig. 22. A, Antifacilitation at crayfish giant motor neuron–phasic flexor muscle junction. Excitatory junctional ("postsynaptic") potentials decline markedly during 1/min stimulation. Open circles, normal physiological medium; filled circles and triangles, high magnesium ion medium (23 and 29 mM Mg Cl2, respectively) to reduce transmitter output. See text (from Bruner and Kennedy, 1970). B, Long-lasting facilitation at motor neuron–leg stretcher muscle junction of *Grapsus grapsus*. Top, EPSPs after 5 min of stimulation at 5/sec; bottom, the same after 40 min of stimulation. This potentiation lasts for over an hour after stimulation is terminated (from Sherman and Atwood, 1971 ; copyright 1970 and 1971 by the American Association for the Advancement of Science).

tion, is illustrated in Fig. 22B (from Sherman and Atwood, 1971). Repetitive stimulation of the single excitatory axon to the leg stretcher muscle of the Bermuda crab, *Grapsus grapsus*, at a rate of five times per second causes an augmentation of junctional potential amplitude. If this stimulation is sufficiently prolonged, as it was in the experiment illustrated, the augmentation persists for at least an hour after the cessation of stimulation.

It is evident that phenomena such as these occurring *within* the central nervous system could be responsible for a variety of the short-term changes of responsiveness discussed in previous pages. Coupled with appropriate circuitry, they could also provide short-term memory for virtually any sort of learning phenomenon. Furthermore, it is often argued—or more often hinted—that similar phenomena might occur over much longer durations at as yet unstudied sites within the central nervous system.

Processes of facilitation, post-tetanic potentiation, and/or antifacilitation occur in greater or lesser degrees at most synapses. Their occurrence at neuromuscular junctions, however, permits them to be studied more easily and intensively than would be the case were they to occur only at synapses buried within the central nervous system. It is of course a presumption that neuromuscular and central synaptic processes are essentially similar, but it is a presumption which has very often proved to be true. Crustacean neuromuscular junctions are by no means unique among nerve–muscle synapses in displaying these phenomena, which have in fact been much more intensively studied at vertebrate junctions. The examples given above are, however, unusual in their considerable longevity and therefore are of special interest to the student of plasticity.

It is generally agreed that effects such as these are due to changes in transmitter output of presynaptic endings and not to variations in postsynaptic responsiveness. However, the mechanisms responsible for these output changes either in Crustacea or in other animals are not understood and surely will not be so until the mechanisms involved in transmitter release are more fully elucidated. There is, however, no dearth of hypotheses each with supporting evidence in its favor. These hypotheses include depletion or recruitment of transmitter available for immediate release, alterations in ionic distributions and fluxes—both actively driven and passive—which affect terminal resting potential or spike size, variations in tonic and phasic amounts of intraterminal calcium ions, differential invasion of spikes into endings, and refractoriness or supernormality of recently used "release sites."

In the particular cases we have described, there is some information germane to the selection of a subset of these hypotheses. Bruner and Kennedy (1970) investigated the effect of increased magnesium ions on the habituation phenomenon which they described to see whether reduction of transmitter output would diminish rate of habituation. They found that the initial, rapid habituation which occurred over the first few trials was unaffected, whereas the subsequent, slowly developing depression which followed was slowed (Fig. 22A). Thus they argued that transmitter depletion is probably not responsible for the first effect, but it might be for the second.

Sherman and Atwood (1971) demonstrated that ouabain, which blocks active sodium transport, greatly potentiated the rate of development of the facilitation phenomenon they described. Furthermore, depletion of external sodium ions prevented development of long-term facilitation. Their interpretation of these observations is that long-term facilitation is due to the accumulation of intraterminal sodium ions during repeated nerve stimulation. It is then argued, following the work of Birks and Cohen (1968), that the intraterminal sodium causes the liberation of intraterminal calcium ions from a molecule that binds them. Since internal calcium ions are believed to directly precipitate transmitter release (see Katz, 1971, for overview), the sodium indirectly facilitates such release. These accounts are obviously tentative and incomplete, but they illustrate the level and kind of analysis possible at neuromuscular synapses.

Neuromuscular synapses have also been used in a preliminary way to look for morphological features which might correlate with profoundly facilitatory vs. antifacilitatory effects. Atwood (1967) has compared the endings of fast-closer axons in the leg of the crab Pachygrapsus to those of excitatory axons which innervate the chronic extensor muscles of the crayfish tail. The former have high initial probability of transmitter output and fatigue rapidly, whereas the latter have low initial output and show good facilitation. The salient characteristics of the nonfacilitating as compared to the facilitating endings are (1) larger diameter of the endings, (2) lower density of synaptic vesicles, (3) larger ratio of specialized synaptic membrane to available vesicles, and (4) more intimate association of synaptic membrane with contractile elements. These differences make some intuitive sense and should encourage further analysis of this kind.

Processes of sensory adaptation also provide interesting models of central change and show that lability occurs in situations where transmitter release is not essentially involved (see Galeano and Chow, 1970).

2. Altered Central Responses

While the analysis of processes of facilitation and antifacilitation is most conveniently done on peripheral synapses, similar phenomena are also very common within the crustacean central nervous system (e.g., see Kennedy and Preston, 1963; Krasne, 1969, and below; Zucker, 1971).

In addition, however, central units show a great variety of more or less persistent alterations in reactivity, firing rate, firing pattern, and so on. The observations of Taylor (1970) discussed previously, in which sensory excitability of an identified interneuron changed, in either direction, for long periods of time (minutes or hours) either spontaneously or in response to transient stimuli, are typical. Kennedy and Preston (1963) have catalogued a number of such phenomena seen incidentally in the course of other studies. It may be anticipated that as the circuits in which various identified units participate are elucidated, the mechanisms of alterations in their behavior will become subject to detailed study.

3. Developmental Analogues.

Processes of development and differentiation of both the nervous system and other organ systems are in many ways analogous to processes of learning and memory. In each case, cellular environment at one time determines cell form or function at later times, in some cases for the life of the organism (Bonner, 1965; Edds, 1967; Jacobson, 1970). That the cellular biology of these "instructive" and "retentive" processes might have important common features would seem to be manifestly apparent.

So far, there has been very little work on any invertebrate comparable to the experimental embryological work which has given us such fascinating glimpses into the nature of development of the vertebrate nervous system (reviewed by Edds, 1967; Jacobson, 1970). However, it is clear from the literature that several investigators perceive that the same features which make Crustacea valuable for neurophysiological analysis give them great potential in the study of the development of the nervous system as well. So far, this perception has led mainly to investigations concerning the specificity and invariance of various features of identified neurons. Thus Stretton and Kravitz (1968) utilized (for the first time) the dye procion yellow to determine the constancy of dendritic branching patterns of the inhibitors to the fast flexor muscles of lobster abdomens. They found that the main features of the branching patterns were quite constant, although there was some "microheterogeneity" in finer branches. Kennedy and his coworkers, utilizing the above method together with electrophysiological techniques, have demonstrated the invariance of a variety of contacts between pairs of identified neurons (Kennedy et al., 1969; Larimer et al., 1971; Davis, 1970) in lobsters and crayfish. Otsuka et al. (1967) have shown the constancy of aspects of the biochemical architecture of identified units of lobster abdominal ganglia. Atwood and Bittner (1971) have studied the distribution of temporal facilitation among the multiple terminals of crustacean motor neurons. A single excitor and one of two inhibitor neurons branch extensively and contact each muscle fiber of the crayfish claw opener and crab walking leg stretcher muscle a number of times. Atwood and Bittner found that all of the terminals, both excitatory and inhibitory, on a given muscle fiber were of the same type with regard to the extent to which they showed temporal facilitation of transmitter output.

These various "structural" studies immediately pose the problem of what determines the characteristics in question. What stimulus switches a developing neuron into one that produces a particular transmitter? On what does the maintenance of this characteristic depend? What determines that a particular neuron shall branch in a particular way? What interactions between nerve cells and muscle fibers lead to all of the endings on a given muscle fiber having a characteristic facilitation index? And so on. If one believes that learning is a matter of development of new connections, alterations in efficacy of old ones, or changes in metabolic activities, the answers to the above developmental questions become crucial guides to formulating hypotheses about learning mechanisms.

An interesting developmental question intimately related to adult nervous plasticity is that of the extent to which maintenance of particular differentiated characteristics depends on the continuing influence of contiguous neurons either via patterns of input or via chemical factors. A crude way to get at this question in vertebrates has been to study the long-term effects of removing or disconnecting parts of the system. Such experiments have led to the discovery of disuse supersensitivity, collateral spouting, fiber by fiber specificity of regeneration, recovery of function, and so on. Few such experiments have been done on invertebrate central nervous systems.

An interesting exception are experiments on the gradual development of central compensation for the removal of sense organs. Many such experiments have been done on insects (Carthy, 1958). An interesting example in decapod Crustacea is the work of Schöne (1954, 1961a) on statolith removal. If statoliths are removed on one side of the body, an asymmetry of posture is produced which wanes over several days. If the remaining statolith is then removed, the converse asymmetry appears, indicating the existence of a central compensation process (Fig. 23).

Wine (1971) has also shown that after severing of the crayfish nerve cord between thorax and abdomen behavioral and electrophysiological reflexes of the abdomen which were previously either absent or very weak

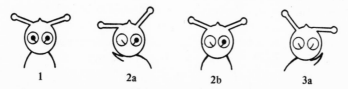

Fig. 23. Posture after operations on statocysts. The diagrams show eyestalk and leg positions when the body is held in its normal orientation. 1, Normal animal. 2a, After removal of left statolith. 2b, Same after several days. 3a, After removal of remaining statolith. (From Schöne, 1961a.)

gradually gather strength over a period of several weeks. There are some difficulties in the interpretation of this result, but it suggests that the functional properties of crustacean neurons depend on a continuing interaction of elements within the nervous system as a whole.

B. Method II: Circuit Analysis

Globally speaking, the second approach to discovery of the neural substrate of learning is to start with known cases of behavioral plasticity, analyze the neuronal circuitry of the behavior pattern in question, and finally see how the functional properties of the circuitry alter when behavior changes. It is, of course, this approach for which invertebrates in general and Crustacea in particular are so well suited, because their conservative use of neurons actually makes it feasible to work out the circuitry of some behavior patterns in something like a complete manner. Once one knows the circuitry, pinpointing loci of change is relatively straightforward.

1. The Lateral Giant Fiber Escape Reflex

From the point of view of circuit analysis, the lateral giant fiber escape reflex of the crayfish may be the most completely understood behavioral reflex known, aside from the monosynaptic stretch reflexes of the mammals (see Wiersma, 1961; Roberts, 1968; Krasne, 1969; Larimer *et al.*, 1971; Zucker *et al.*, 1971; Wine and Krasne, 1972; Zucker, 1972; and references therein). Behaviorally speaking, this reflex is a rapid abdominal flexion which occurs in response to phasic tactile stimuli to the adbomen of well-rested animals. The flexion propels the animal backward and upward through the water in such a way that it evades further contact with the source of disturbance (Larimer *et al.*, 1971; Wine and Krasne, 1972). For our present purposes, the interest of this reflex is that it is very prone to habituation (or, more neutrally, *failure*). This is shown in Figs. 24 and 25. Such behavioral habituation occurs with stimulus frequencies as low as (and probably lower than) one per 5 min, and recovery occurs over a period of hours, probably being complete after somewhere between 24 and 48 hr of rest.[5]

The overall organization of the reflex is shown in Fig. 26 in schematic form. Only features relevant to our present interest have been included. In brief, sensory fibers enter the ganglion and synapse either directly or indirectly with dendrites of the lateral giant fibers. The lateral giants are themselves a chain of neurons which make highly efficient electrical connections with one another and therefore *functionally* constitute a single long axon with dendritic ramification in each segmental ganglion. The lateral giants in turn synapse with motor neurons to flexor muscles in each abdominal segment. [It was the

[5]The upper limits have not been accurately evaluated.

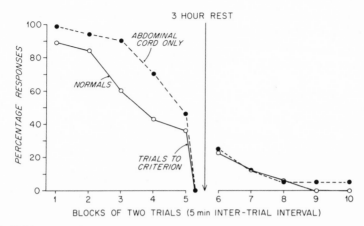

Fig. 24. Habituation of crayfish lateral giant fiber escape reflex. Stimuli (taps to the side of the abdomen) were given at 5-min intervals except during the 3-hr rest and during trials to criterion. Trials to criterion consisted of one tap per minute in blocks of ten separated by 10-min rests until completion of a block without any responses. The "abdominal cord only" group is discussed in the text (data from Wine and Krasne, 1969; and unpublished material).

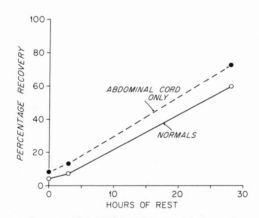

Fig. 25. Time course of recovery from habituation to criterion as shown at the left of Fig. 24. "Percentage recovery" is defined as the percentage of responses occurring in the ten trials after the rest relative to the first ten trials of the experiment. Recovery tests are given serially (data from Wine and Krasne, 1969).

neuromuscular synapses of one of the motor neurons, the so-called motor giants, which were used by Bruner and Kennedy (see above) to illustrate neuromuscular antifacilitation.]

Recordings from intact, chronic preparations show that behavioral habituation is the result of a transmission failure between sensory neurons

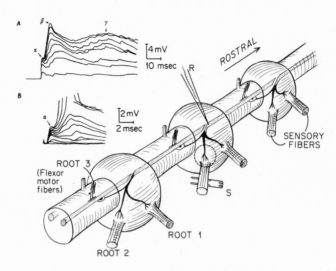

Fig. 26. Schematic diagram of middle portion of crayfish abdominal nerve cord to show general organization of lateral giant fiber escape reflex. Only the lateral giant of the right side is shown. To get the results shown in A and B, electrical shocks were applied at *S* and an intracellular recording electrode was placed at *R*. A and B are from different animals and show superimposed the traces for a series of stimulus intensities. The dashed circle in the center ganglion indicates the area diagrammed in Fig. 27 (A and B from Krasne, 1969).

and lateral giants (Wine and Krasne, 1969). Acute experiments have permitted the analysis to be carried further. Synchronous phasic activation of sensory fibers (by a single electrical shock) causes a compound EPSP to appear at the base of the lateral giant's dendritic tree in the stimulated segment (Fig. 26), and the form of this potential as well as the way that it varies with stimulus strength allows some informed guesses to be made about the pattern of connection between the sensory fibers and the lateral giants (Krasne, 1969). The very short latency (as little as 0.4 msec) of the first (α) component suggests that it is monosynaptic and indeed perhaps electrical, while the latency of the second (β) component, which may be as little as 1.3 msec, suggests that it is probably disynaptic if the junctions are chemical ones. The pattern which might be thus envisaged is shown in Fig. 27. Since the behavioral reflex is labile, and since it is the β-component which is responsible for lateral giant firing, both in acute preparations (Krasne, 1969) and in intact animals (Wine and Krasne, 1972), we would expect this component itself to show some kind of lability. Experiments with stimulus repetition do in fact show (Krasne, 1969) that whereas the α-pathway is very stable, the β-pathway is subject to pronounced antifacilitation (Fig. 28). Furthermore, the failure of the EPSP during stimulus repetition occurs in a saltatory manner (Fig. 29A), which suggests that failure may be occurring at the first tier of hypothetical synapses in Fig. 27 and causing interneurons to drop out in an all-or-none fashion.

Recently, it has been found (Zucker *et al.*, 1971; Zucker, 1972) that several large neurons which run in the connectives near their ventral margin are in fact interneurons of the sort envisaged above, and the study of these neurons has allowed verification and further elaboration of the above picture, based of course on the assumption that the interneurons in question are typical of the interneurons of the reflex arc as a whole. Such study proves directly that it is indeed the first synapses and not the second which are labile and shows, furthermore, that the labile junctions are apparently chemical, whereas the stable junctions, which have very short synaptic delays, are electrical (Zucker *et al.*, 1971; Zucker, 1972). Recordings from the dendrites of the interneurons show that unitary excitatory postsynaptic potentials decline gradually during stimulus repetition (Fig. 29B), and statistical analysis

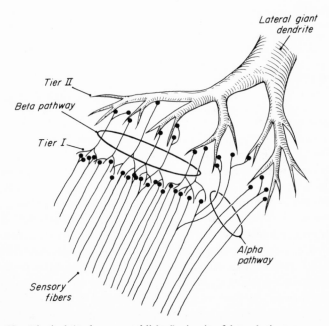

Fig. 27. Hypothetical (and now established) circuit of lateral giant escape reflex. The region where the circuitry operates is shown as a dashed circle in Fig. 26. It is believed that most of the synapses onto the lateral giants are electrical and temporally stable, whereas Tier I synapses of the β-pathway are chemical and show antifacilitation. Some of the β-pathway interneurons send branches into the connectives between ganglia and make homologous contacts in other abdominal ganglia (further details of the pathway are described in Zucker *et al.*, 1971).

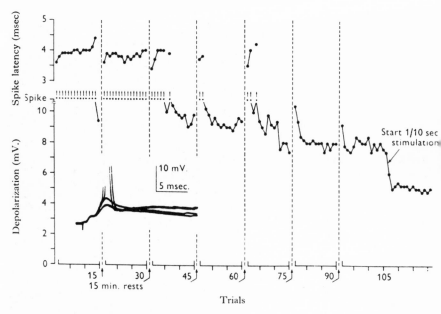

Fig. 28. Response failure as a result of stimulation once per minute. Blocks of 15 shocks at 1/min were alternated with 15-min periods of rest. The size of the β-component is plotted on trials without a spike; spike latency is plotted on trials with one. The oscilloscope traces from trials 61, 62, 63, and 75 have been superimposed in the lower left of the figure (from Krasne, 1969).

of the variability of unitary EPSP size before and after reflex fatigue suggests that the decrease is due to a reduction in the number of quanta of transmitter released rather than to desensitization of the postsynaptic cell (Zucker, 1972).

There is no indication in these experiments that habituation might be due to either pre- or postsynaptic inhibitory influences, and picrotoxin, which poisons all *known* forms of crayfish inhibition, does not prevent or slow the course of reflex failure in this system (Krasne and Roberts, 1967). Additional circumstantial evidence that reflex failure in these preparations is due to an intrinsic process such as transmitter depletion rather than to an extrinsic gating follows from the fact that during a tail-flip response transmission between sensory fibers and the interneurons of the β-pathway is inhibited and also *protected* from habituation (Fig. 30). The most parsimonious interpretation of the parallel time courses of this inhibition and protection is that during a tail flip the afferent terminals which feed the lateral giant reflex are presynaptically inhibited, thus preserving transmitter substance and preventing all subsequent excitatory actions. Of course, more elaborate explanations consistent with some form of extrinsic gating could be conceived.

Thus appears that this case of behavioral habituation has been local-

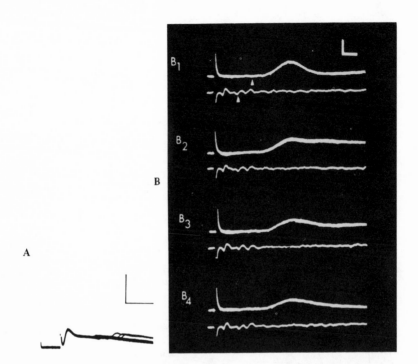

Fig. 29. A, Saltatory failure of the β-component. The recording site is the base of the lateral giant dendrite. Stimulus strength was adjusted so that only a single interneuron of the β-pathway fired. Repetitions of this stimulus are shown on superimposed sweeps. Calibration 5 mv and 5 msec (from Krasne, 1969). B, Gradual failure of a unitary EPSP in a β-pathway interneuron. B_1–B_4 show responses on alternate trials at $1/2$ sec. Upper traces, EPSP; lower traces, sensory nerve activity. Calibration 2 mv and 2 msec (from Zucker *et al.*, 1971; copyright 1971 by the American Association for the Advancement of Science).

ized to a single synaptic layer and has been shown (at least tentatively) to be due to a presynaptic process.

An interesting point which emerges from this conception is that the malleable synapses are the chemical and not the electrical ones. Alterations in efficacy of synaptic transmission could in principle be brought about either by alterations in chemical transmitter output or effectiveness or by changes in critical firing thresholds of postsynaptic cells. The fact that the one means rather than the other has been utilized may be a point of general significance.

Pretty as it is, the above story may be misleadingly simple, for when one compares the time courses of habituation and recovery in the acute experiments described above to those in intact animals, there is a striking discrepancy (compare Fig. 31 to Fig. 25 above).

A fact which at one time offered hope of explaining this discrepancy is

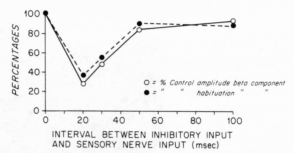

Fig. 30. Relationship between inhibition (open circles) and protection from habituation (filled circles) of the β-component of the lateral giant escape reflex. The habituation series was ten stimuli at 1/5 sec (unpublished data of Joan Bryan from the author's laboratory).

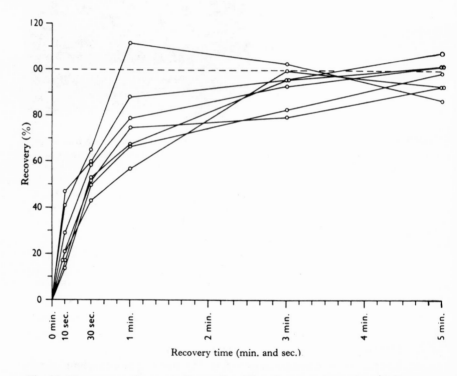

Fig. 31. Time course of recovery from reflex failure in acute preparations. Failure was induced by ten stimuli at 5-sec intervals. Each point shown is the increment in β-component amplitude occurring during a rest of the stated duration and is given as a percentage of the difference between the depolarizations evoked on the first and last trials of the preceding conditioning bout. The lines connect the points from single animals (from Krasne, 1969).

that in intact, reasonably healthy animals there is a descending inhibitory system operative which is capable of modulating the excitability of the lateral giant reflex (Wine and Krasne, 1969; and Fig. 32). A reasonable hypothesis was that it is the operation of this system which makes the difference between the short-lived habituation of acute preparations and the relatively long-term habituation of normal animals. However, this seems not to be the case, for chronic preparations whose abdominal nerve cords are separated from more rostral parts show no sign of having an inhibitory system operative and yet show patterns of habituation and recovery which are virtually identical to those of normals (Wine and Krasne, 1969; and Figs. 24 and 25).

There are many technical differences between acute and chronic habituation experiments, and it may be only a matter of time before the crucial variables are isolated. However, one possibility which must be faced is that excitability levels, extent and character of spontaneous activity, and so on are abnormal enough in acute preparations to make normal behavior impossible (see Farel and Krasne, 1972). After all, one cannot expect a crayfish tail that is opened, isolated, pinned down, and has Ringer solution flowing over it to behave in an entirely normal fashion. And as we use invertebrates to study progressively more subtle behavior, we may find it increasingly difficult to get normal behavior out of physiological preparations.

More or less similar conclusions to those drawn for the lateral giant escape reflex have been come to by investigators studying various sorts of habituation in the gastropod mollusc *Aplysia* (Bruner and Tauc, 1966; Castellucci *et al.*, 1970), and it is entirely possible (Horn, 1967) that habituation of reflexes generally, or as Sherrington called it, "reflex fatigue," is indeed

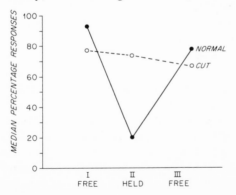

Fig. 32. Inhibition of the lateral giant escape reflex. Animals were given single-escape reflex trials while free, while held by the carapace out of water, and while free again. The reflex of normal animals is clearly suppressed when they are held. The "cut" group, which is not affected by being held (either by thorax or by abdomen), contains animals whose nerve cords were sectioned either at the abdominal–thoracic junction or in the middle of the thorax. The three trials within a test were given about 1 min apart (unpublished data of Sun Hee Lee and Jeff Wine from the author's laboratory).

due to what might be called "synaptic fatigue." An interesting case of habituation which on the face of it seems as though it might involve active suppression of responses is the habituation of the defensive claw-opening response that is elicited in crayfish when their backs are stroked. This habituation is said to be prevented if the inhibitor axon to the opener muscle of the claw is cut (von Buddenbrock, 1953; Schöne, 1961a). It could be that this operation merely lowers the threshold at which muscular excitation produces a behavioral response so that a reduced (habituated) excitatory output is nevertheless effective in causing movement. But the *possibility* of inhibition of movement is interesting and should be investigated.

2. Preparations for Associative Learning

The above example is nice enough, but habituation is not associative learning. While we have seen ample evidence that Crustacea learn, the experimental situations that have been employed have not been of a kind that lead easily to physiological analysis. There are, however, a few preparations that look as though they might well be profitably exploited to tell us something about learning in the sense of forming new associations.

In Chapter 8 of this volume, devoted to the insects, there is an extended discussion of a training procedure invented by Horridge in which one of an animal's legs is allowed to dangle over an electrical conductor in such a way that when the animal relaxes the leg sufficiently, it touches the conductor and receives a shock. Insects with and without brains, spinal rats (Buerger and Fennessy, 1971), and spinal frogs (Horn and Horn, 1969; Farel and Buerger, 1972) all learn to keep their legs lifted under such circumstances. The beauty of the procedure, at least in insects, is that the single hemiganglion which innervates the withdrawing leg is sufficient for the learning (Chapter 8, this volume). Thus the problem of physiological analysis is greatly simplified. The Crustacea would seem to be especially favorable for the analysis of this sort of effect because of our growing knowledge of their motor control systems. In particular, the control of specific coordinated movements by identified command fibers greatly truncates the problem of circuit analysis.

Probably the most inherently promising preparation for such experiments, which unfortunately has not been exploited, is the system for tonic control of abdominal posture in the crayfish. Kennedy and his coworkers (Kennedy et al., 1966a,b; Evoy and Kennedy, 1967) have shown that each of a whole range of slow abdominal movements is under the control of a separate command fiber. Thus there are fibers which elicit flexions or extensions of various parts of the abdomen, fibers which bring about slower or faster movements, fibers which cause abdominal movements only, others which do so in conjunction with motions of swimmerets or uropods, and so on. The obvious experiment is to see whether the posture of the crayfish tail can be controlled by Horridge's sort of procedure. If it can, one would then be in a very good position to carry through a detailed analysis.

While the Horridge procedure has not been tried in the above situation, it has apparently been utilized successfully to control leg position in several kinds of crabs (Donald Rafuse, personal communication). If, for example, *Hemigrapsus nudus* or *Cancer productus* with all but one leg removed is suspended over a conducting plate and shocked whenever its leg touches the plate, the animal rapidly learns to keep the leg raised. Experiments with control animals that receive like shocks but cannot control their arrival seem to show that true learning, although probably of short duration, is involved. Rafuse has shown that the supraesophageal ganglion is not required for this learning and has identified a cluster of cells in the subesophageal ganglion which seems to contain command units for leg movements. These units send axons to the immediate vicinity of appropriate motor neurons and have dendritic ramifications in the immediate vicinity of sensory neurons from the legs. Finally, Rafuse has stained for perinuclear RNA (Cohen and Jacklet, 1965) and studied uptake of radioactive uridine into the somas of his command cluster units and found changes which occur in trained animals but not in yoked controls (personal communication). If these very interesting results stand up to detailed scrutiny, we obviously have here a very exciting preparation.

C. Method III: Holistic Procedures

That certain kinds of very interesting information about the learning process can be obtained without the benefit of detailed analysis of neural activity is illustrated by experiments on the effects of treatments such as electroconvulsive shock and protein synthesis inhibitors on memory in vertebrates. While the interpretations of many, if not all, of these experiments are subjects of hot controversy, it cannot be denied that such methods at least *seem* to be capable of providing us with important insights about the learning process.

While all of these kinds of experimentation could, of course, be carried out on Crustacea with some profit, I wish to propose here a line of holistic experimentation for which Crustacea appear to be particularly suited. A major question about the cellular bases of learning revolves around the extent to which interactions between cytoplasm and nuclear genome are required, on the one hand, for the acquisition of learned behavior and, on the other, for its retention. All invertebrates provide interesting material for the study of this question because neuron somas are set aside from the parts of the cell which are most obviously involved in information processing, and this makes it possible to examine the effects on learning or retention of removing the cell body without totally disrupting the activities of the neuron. This is especially true because the neuron somas occur in relatively neat clusters at the periphery of the nervous system. In 1896, von Bethe performed in a few cases the very difficult feat of (1) isolating the neuropile of the second antenna of *Carcinus* from the remainder of the nervous system and (2) removing all of its cell bodies. In the best of the several cases where he performed the operation successfully, both reflex and spontaneous movement of the antenna remained more or less normal for some 3 days after the operation; after that, he unfortunately sacrificed the animal to carry out histological tests, thinking he had perhaps done an unsatisfactory operation or gotten his animals mixed up. In other cases, the results were not so spectacular, but the point that crustacean reflexes can remain quite intact without cell bodies is nevertheless made.

Various sorts of short-term learning appear to occur normally after removal of cell bodies. Thus temporal facilitation as well as antifacilitation (or "habituation") at nerve–muscle junctions or sensory interneuron junctions (all of which were described above) is not conspicuously affected by acute removal of cell bodies, which for reasons of convenience is done routinely in many experiments. Furthermore, Hoyle (verbal communication) claims that he is able to satisfactorily carry out Horridge's training procedures on an insect ganglion whose cell bodies have been acutely removed. However, these short-term learning phenomena are not the sort which would be most expected to involve access to the genome. The most informative experiments will therefore be ones in which the role of the cell bodies in long-term learning and retention is investigated.

It is here that crustacean material may have special virtues, for whereas the processes of most neurons begin to degenerate shortly after removal of their somas, there is growing evidence to suggest that both propagation and synaptic events in crustacean neurons may remain normally functional for weeks or months after they have lost their cell bodies. Hoy *et al.* (1967) have demonstrated this for motor neurons to the crayfish claw, and Wine (1972) has shown it for the median giant fibers of the crayfish nerve cord; in both cases, synaptic transmission remained normal for more than 100 days after the operation! While to some extent these may be special cases, it does appear likely that general survival will turn out to be long enough to permit tests for "long-term" learning and retention.

Of course, the problem with actually carrying out the desired sorts of investigation is that it is by no means feasible to remove all the somas of the brain. However, if discrete neuropile regions could be shown to be critical to certain learning tasks, a first approximation to the desired sort of experiment might be possible. There have been hardly any studies of the effects of local central nervous system damage on learning in Crustacea, an unfortunate gap in the literature in its own right, but a study on the crayfish

Procambarus by Eisenstein and Mill (1965) in fact suggests a possible preparation. Their animals were given experience running along the floor of their aquaria toward a striped target adjacent to a bit of Gerber's strained beef located upstream. After some days of this, the animals were tested with the target alone. Normal animals learned to recognize the target. Animals with one eyestalk removed did not; this was not due to impaired vision *per se,* since animals with an eyestalk covered instead of removed did learn. These results are shown in Fig. 33. The eyestalks of the crayfish are not solely visual and in fact contain important association areas including (as Wiersma, 1970, has pertinently pointed out) a probable homologue of the corpora pedunculata (Bullock and Horridge, 1965; Seabrook and Nesbitt, 1966). Perhaps visual learning depends on the contralateral "corpus pedunculatum"; perhaps two corpora pedunculata are better than one. In any case, study of the effects of extirpation (before or after learning) of the cell bodies of this neuropile area rather than of the neuropile itself would be very interresting for a patient investigator to attempt.

A somewhat simpler line of research bearing on similar issues would be an investigation of consolidation time and the effects of temperature on consolidation in Crustacea or other invertebrates. If the genome is critically involved in long-term learning, one might expect the spatial separation of the soma from dendrites, spike-initiating zones, and so on, to lead to differences in parameters of consolidation in invertebrate as compared to cold-blooded vertebrate animals. In particular, cooling, which slows axoplasmic flow (Fernandez *et al.,* 1970), might have especially marked effects on invertebrates.

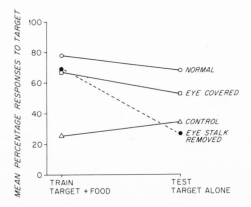

Fig. 33. Effect of eyestalk removal on approach response learning in *Procambarus clarkii.* Whereas the "normal," "eye covered," and "eye stalk removed" groups all show approach responses when target and food are presented, the last group shows deterioration to control levels when food is not present, indicating that it has not associated the target with food. The control group received unpaired presentations of target and food during training (data from Eisenstein and Mill, 1965).

The effects of pharmacological agents which poison axoplasmic flow would also be of great interest in this context.

VII. CONCLUDING REMARKS

In summation, we may say that it is quite certain that many crustaceans can learn, and it is reasonably clear that this ability is utilized in the animals' natural lives. Most formal experiment son learning have been mainly demonstrations that this or that animal could learn; further progress will now require study of the detailed nature of various animals' abilities and their relationship both to the animals' ways of life and to relevant features of their nervous systems. The diversity of the Crustacea makes them ideal for such investigations, yet their potential is virtually untapped. Finally, it has been emphasized that crustacean nervous systems have many features which will aid the physiological analysis of learning. There are several lines of very interesting research taking advantage of these properties. However, as yet we know of no case of *learning* in the vernacular sense of that word (i.e., long-term association) which shows much promise of being subject to the sort of detailed physiological analysis for which crustacean nervous systems offer special advantages. The development of such preparations, if it is presently possible, will require highly creative psychological investigations by individuals who understand the possibilities and limitations of present neurophysiological methods and who are steeped in the known neurology of the animals they study.

REFERENCES

Agar, W. E., 1927, The regulation of behavior in water-mites and some other arthropods, *J. Comp. Psychol., 7,* 39–74.
Allee, W. C., and Douglis, M. B., 1945, A dominance order in the hermit crab, *Pagurus longicarpus* Say, *Ecology, 26,* 411–412.
Amsel, A., 1958, The role of frustrative non-reward in non-continuous reward situations, *Psychol. Bull., 55,* 102–119.
Appellöf, A., 1909, Untersuchungen über den Hommer, *Bergens Museums Shrifter (N.S.), 1,* 2–78.
Applewhite, P. B., and Morowitz, H. J., 1966, The micrometazoa as model systems for studying the physiology of memory, *Yale J. Biol. Med., 39,* 90–105.
Arechiga, H., and Wiersma, C. A. G., 1969, Circadian rhythms of responsiveness in crayfish visual units, *J. Neurobiol., 1,* 71–85.
Atwood, H. L., 1967, Crustacean neuromuscular mechanisms, *Am. Zoologist, 7,* 527–551.
Atwood, H. L., and Bittner, G. D., 1971, Matching of excitatory and inhibitory inputs to crustacean muscle fibers, *J. Neurophysiol., 34,* 157–170.
Atwood, H. L., and Wiersma, C. A. G., 1967, Command interneurons in the crayfish central nervous system, *J. Exptl. Biol., 42,* 249–261.

Bainbridge, R., 1961, Migrations, in "The Physiology of Crustacea" (T. H. Waterman, ed.), Vol. II, pp. 431–455, Academic Press, New York.

Barnes, R. D., 1968, "Invertebrate Zoology," Saunders, Philadelphia.

Barnwell, F. H., 1963, Observations on daily and tidal rhythms in some fiddler crabs from equatorial Brazil, Biol. Bull., 125, 399–415.

Barnwell, F. H., 1968, The role of rhythmic systems in the adaptation of fiddler crabs in the intertidal zone, Am. Zoologist, 8, 569–583.

Bethe, A., 1897, Vergleichende Untersuchungen über die Funktionen des Centralnervensystems der Arthropoden, Arch. Ges. Physiol. Pflügers, 68, 449–545.

Bethe, A., 1898, Das Centralnervensystem von Carcinus maenas. II, Arch. Mikroskop. Anat., 51, 382–452.

Birks, R. I., and Cohen, M. W., 1968, The action of sodium pump inhibitors on neuromuscular transmission, Proc. Roy. Soc. Ser. B, 170, 381–399.

Bitterman, M. E., 1965, Phyletic differences in learning, Am. Psychologist, 20, 396–410.

Blees, G. H. J., 1919, Phototropisme et expérience chez la Daphnie, Arch. Nierl. Physiol., 3, 279–306.

Bock, A., 1942, Über das Lernvermögen bei Asseln, Z. Vergl. Physiol., 29, 595–637.

Bonner, J. F., 1965, "The Molecular Biology of Development," Oxford University Press, New York.

Borradaile, L. A., Potts, F. A., Eastham, L. E. S., Saunders, J. T., and Kerkut, G. A., 1958, "The Invertebrata," Cambridge University Press, Cambridge.

Bovbjerg, R. V., 1953, Dominance order in the crayfish, Orconectes virilis (Hagen), Physiol. Zool., 26, 173–178.

Brereton, J. LeG., 1957, The distribution of woodland isopods, Oikos, 8, 85–106.

Brown, F. A., 1961, Physiological rhythms, in "The Physiology of Crustacea" (T. H. Waterman, ed.), Vol. II, pp. 401–430, Academic Press, New York.

Bruner, J., and Kennedy, D., 1970, Habituation: Occurrence at a neuromuscular junction, Science, 169, 92–94.

Bruner, J., and Tauc, L., 1966, Habituation at the synaptic level in Aplysia, Nature, 210, 37.

Buchsbaum, R., and Milne, L. J., 1960, "The Lower Animals: Living Invertebrates of the World," Doubleday, Garden City, N.Y.

Buerger, A. A., and Fennessy, A., 1971, Long-term alteration of leg position due to shock avoidance by spinal rats, Exptl. Neurol., 30, 195–211.

Bullock, T. H., and Horridge, G. A., 1965, "Structure and Function in the Nervous Systems of Invertebrates," Freeman, San Francisco.

Capretta, P. J., and Rea, R., 1967, Discrimination reversal learning in the crayfish, Anim. Behav., 15, 6–7.

Carthy, J. D., 1958, "An Introduction to the Behavior of Invertebrates," Allen and Unwin, London.

Castellucci, V., Pinsker, H., Kupfermann, I., and Kandel, E. R., 1970, Neuronal mechanisms of habituation and dishabituation of the gill withdrawal reflex in Aplysia, Science, 167, 1745–1748.

Chow, K. L., and Leiman, A. L., The photo-sensitive organs of crayfish and brightness learning, Personal communication and in preparation.

Cohen, M. J., and Jacklet, J. W., 1965, Neurons of insects: RNA changes during injury and regeneration, Science, 148, 1237–1239.

Costa, H. H., 1966, Responses of Gammarus pulex (L.) to modified environment. I. Reactions to toxic substances, Crustaceana, 11, 245–255.

Cowles, R. P., 1908, Habits, reactions, and associations in Ocypoda arenaria, Papers Tortugas Lab. Carnegie Inst. Wash., 2, 1–41.

Crane, J., 1958, Aspects of social behavior in fiddler crabs, with special reference to *Uca maracoani* (Latreille), *Zoologica, 43,* 113–130.

Creaser, E. P., and Travis, D., 1950, Evidence of a homing instinct in the Bermuda spiny lobster, *Science, 112,* 169–170.

Datta, L. G., Milstein, S., and Bitterman, M. E., 1960, Habit reversal in the crab, *J. Comp. Physiol. Psychol., 53,* 275–278.

Daumer, K., Jander, R., and Waterman, T. H., 1963, Orientation of the ghost-crab *Ocypode* in polarized light, *Z. Vergl. Physiol., 47,* 56–76.

Davis, W. J., 1970, Motoneuron morphology and synaptic contacts: Determination by intracellular dye injection, *Science, 168,* 1358–1360.

Dembrowska, W. S., 1926, Study on habits of the crab *Dromia vulgaris* M. E., *Biol. Bull., 50,* 163–178.

Doflein, F., 1910, Lebensgewohnheiten und Anpassungen bei dekapoden Krebsen, *in* "Festschrift zum 60 Geburtstag Richard Hertwigs," Vol. 3, pp. 215–292, Fischer, Jena.

Drzewina, A., 1908, Les réactions adaptives des Crabes, *Bull. Inst. Gen. Psychol., 8,* 235–256.

Drzewina, A., 1910, Création d'associations sensorielles chez les crustacés, *Compt. Rend. Soc. Biol. Paris, 68,* 573–575.

Dudel, J., and Kuffler, S. W., 1961, Mechanism of facilitation at the crayfish neuromuscular junction, *J. Physiol., 155,* 530–542.

Edds, M. V., Jr., 1967, Neuronal specificity in neurogenesis, *in* "The Neurosciences, A Study Program" (G. C. Quarton, T. Melnechuck, and F. O. Schmitt, eds.), pp. 230–240, and Rockefeller University Press, New York.

Eisenstein, E. M., and Mill, P. J., 1965, Role of the optic ganglia in learning in the crayfish *Procambarus clarkii* (Girard), *Anim. Behav., 13,* 561–565.

Evoy, W. H., and Kennedy, D., 1967, The central nervous organization underlying control of antagonistic muscles in the crayfish. I. Types of command fibers, *J. Exptl. Zool., 165(2),* 223–248.

Farel, P. B., Buerger, A. A., 1972, Instrumental conditioning of leg position in chronic spinal frog: Before and after sciatic section, *Brain Res., 47,* 345–351.

Farel, P. B., and Krasne, F. B., 1972, Maintenance of habituation by infrequent stimulation, *Physiol. Behav., 8,* 783–785.

Fernandez, H. L., Huneeus, F. C., and Davison, P. F., 1970, Studies on the mechanism of axoplasmic transport in the crayfish cord, *J. Neurobiol., 1,* 395–409.

Fink, H. K., 1941, Deconditioning of the "fright reflex" in the hermit crab, *Pagurus longicarpus, J. Comp. Psychol., 32,* 33–39.

Fraenkel, G. S., and Gunn, D. L., 1961, "The Orientation of Animals," Dover, New York.

Galeano, C., and Chow, K. L., 1970, Response of caudal photoreceptor of crayfish to continuous and intermittent photic stimulation, *Can. J. Physiol. Pharmacol., 49,* 699–706.

Garcia, J., McGowan, B. K., Ervin, F. R., and Koelling, R. A., 1968, Cues: Their relative effectiveness as a function of the reinforcer, *Science, 160,* 794–795.

Gilhousen, H. C., 1927, The use of the vision and of the antennae in the learning of crayfish, *Univ. Calif. (Berkeley) Publ. Physiol., 7,* 73–89.

Gregory, R. L., 1966, "Eye and Brain," McGraw-Hill, New York.

Gwilliam, G. F., 1966, The mechanism of the shadow reflex in Cirripedia. II. Photoreceptor cell response, second-order responses and motor cell output, *Biol. Bull., 131,* 244–256.

Harker, J. E., 1964, "The Physiology of Diurnal Rhythms," Cambridge University Press, Cambridge.

Harless, M. E., 1967, Successive reversals of a position response in isopods, *Psychon. Sci., 9,* 123–124.

Harris, J. E., 1963, The role of endogenous rhythms in vertical migration, *J. Marine Biol. Ass. G.B., 43,* 153–166.

Hazlett, B. A., 1966, Temporary alteration of the behavioral repertoire of a hermit crab, *Nature, 210,* 1169–1170.

Hazlett, B. A., 1969, "Individual" recognition and agonistic behavior in *Pagurus bernhardus, Nature, 222,* 268–269.

Hazlett, B. A., 1971, Influence of rearing conditions on initial shell entering behavior of a hermit crab (Decapoda Paguridea), *Crustaceana, 20(2),* 167–170.

Hazlett, B. A., and Provenzano, A. J., 1965, Development of behavior in laboratory reared hermit crabs, *Bull. Marine Sci., 15,* 616–633.

Held, R., and Hein, A., 1963, Movement-produced stimulation in the development of visually guided behavior, *J. Comp. Physiol. Psychol., 56,* 872–876.

Herrnkind, W. F., 1968, Adaptive visually-directed orientation in *Uca pugilator, Am. Zoologist, 8,* 583–598.

Hertz, M., 1932, Verhalten des Einsiedlerkrebses *Clibanarius misanthropicus* gegenüber verschiedener Gehäuseformen, *Z. Vergl. Physiol., 18,* 597–621.

Hertz, M., 1933, Über figurale Intensitäten und Qualitäten in der optischen Wahrenehmung der Biene, *Biol. Zbl., 53,* 10–40.

Hertz, M., 1934, Zur Physiologie des Formen und Bewegungssehens. III. Figurale Unterscheidung und reziproke Dressuren bei der Biene, *Z. Vergl. Physiol., 21,* 604–615.

Hodos, W., 1970, Evolutionary interpretation of neural and behavioral studies of living vertebrates, *in* "The Neurosciences. Second Study Program" (G. C. Quarton, T. Melnechuck, and G. Adelman, eds.), pp. 26–39, Rockefeller University Press, New York.

Horn, A. L. D., and Horn, G., 1969, Modification of leg flexion in response to repeated stimulation in a spinal amphibian (*Xenopus mullerei*), *Anim. Behav., 17,* 618–623.

Horn, G., 1967, Neuronal mechanisms of habituation, *Nature, 215,* 707–711.

Horridge, G. A., 1966a, Optokinetic memory in the crab, *Carcinus, J. Exptl. Biol., 44,* 233–245.

Horridge, G. A., 1966b, Optokinetic responses of the crab, *Carcinus,* to a single moving light, *J. Exptl. Biol., 44,* 263–274.

Horridge, G. A., 1966c, Direct response of the crab, *Carcinus,* to the movement of the sun, *J. Exptl. Biol., 44,* 275–283.

Horridge, G. A., 1968, "Interneurons," Freeman, San Francisco.

Hoy, R., Bittner, G., and Kennedy, D., 1967, Regeneration in crustacean motoneurons: Evidence for axonal fusion, *Science, 156,* 251–252.

Hughes, G. M., 1965, Neuronal pathways in the insect central nervous system, *in* "The Physiology of the Insect Central Nervous System" (J. E. Treherne and J. W. L. Beament, eds.), pp. 79–112, Academic Press, New York.

Hughes, R. N., 1966, Some observations of correcting behavior in woodlice (*Porcellio scaber*), *Anim. Behav., 14,* 319.

Iwahara, S., 1963, Inhibition vs. thigmotropism vs. centrifugal swing as determinates of the initial turn alternation phenomenon in *Armadillidium vulgare, Ann. Anim. Psychol. Tokyo, 13,* 1–15.

Jacobson, M., 1970, Development, specification and diversification of neuronal connections, *in* "The Neurosciences. Second Study Program" (G. C. Quarton, T. Melnechuck, and G. Adelman, eds.), pp. 116–129, Rockefeller University Press, New York.

Katz, B., 1971, Quantal mechanism of neural transmitter release, *Science, 173,* 123–126.

Kennedy, D., and Preston, J. B., 1963, Post activation changes in excitability and spontaneous firing of crustacean interneurons, *Comp. Biochem. Physiol., 8,* 173–179.

Kennedy, D., Evoy, W. H., and Fields, H. L., 1966a, The unit basis of some crustacean reflexes, *in* "Nervous and Hormonal Mechanisms of Integration," Vol. 20, pp. 75–109, *Symp. Soc. Exptl. Biol.,* Academic Press, New York.

Kennedy, D., Evoy, W. H., and Hanawalt, J. F., 1966b, Release of coordinated behavior in crayfish by single central neurons, *Science, 154,* 917.

Kennedy, D., Selverston, A. I., and Remler, M. P., 1969, Analysis of restricted neural networks, *Science, 164,* 1488–1496.

Krasne, F. B., 1969, Excitation and habituation of the crayfish excape reflex: The depolarizing response in lateral giant fibers of the isolated abdomen, *J. Exptl. Biol., 50,* 29–46.

Krasne, F. B., and Roberts, A. M., 1967, Habituation of the crayfish escape response during release from inhibition induced by picrotoxin, *Nature, 215,* 769–770.

Krasne, F. B., and Woodsmall, K. S., 1969, Waning of the crayfish escape response as a result of repeated stimulation, *Anim. Behav., 17,* 416–424.

Kühl, H., 1933, Die Fortbewegung der Schwimmkrabben mit Bezug auf die Plastizität des Nervensystems, *Z. Vergl. Physiol., 19,* 489–521.

Kupfermann, I., 1966, Turn alternation in the pill bug (*Armadillidium vulgare*), *Anim. Behav., 14,* 68–72.

Lagerspetz, K. Y. H., and Kivivuori, L., 1970, The rate and retention of the habituation of the shadow reflex in *Balanus improvisus* (Cirripedia), *Anim. Behav., 18,* 616–620.

Larimer, J., Eggleston, A., Masukawa, L., and Kennedy, D., 1971, The different connections and motor outputs of lateral and medial giant fibers in the crayfish, *J. Exptl. Biol., 54,* 391–402.

Lindauer, M., 1961, "Communication Among Social Bees," Havard University Press, Cambridge, Mass.

Lindberg, R. G., 1955, Growth, population dynamics, and field behavior in the spiny lobster *Panulirus interruptus* (Randall), *Univ. Calif. Publ. Zool., 59,* 157–247.

Lowe, M. E., 1956, Dominance–subordinance relationships in *Cambarellus shufeldtii, Tulane Stud. Zool., 4,* 139–170.

Luther, W., 1930, Versuche über die Chemorezeptoren der Brachyuran, *Z. Vergl. Physiol., 12,* 177–205.

MacGinitie, G. E., and MacGinitie, N., 1949, "Natural History of Marine Animals," McGraw-Hill, New York.

Maynard, D. M., 1955, Activity in a crustacean ganglion. II. Pattern and interaction in burst formation, *Biol. Bull., 109,* 420–436.

Mikhailoff, S., 1920, Expérience réflexologique—L'activité neuro-psychique (formation des réflexes associés) est-elle possible sans l'écorce cérébrale? (Première communication préliminaire). Analyse de l'état actuel de la question et expériences nouvelles sur *Pagurus striatus, Bull. Inst. Océanograph. Monaco,* No. 375.

Mikhailoff, S., 1922, Expérience réflexologique: Expériences nouvelles sur *Pagurus striatus Bull. Inst. Océanograph. Monaco,* No. 418.

Mikhailoff, S., 1923, Expérience réflexologique: Expériences nouvelles sur *Pagurus striatus, Leander xiphiaas et treillianus, Bull, Inst. Océanograph. Monaco,* No. 422.

Milne, L. J., and Milne, M., 1961, Scanning movements of the stalked compound eyes in crustaceans of the order Stomatopoda, *in* "Progress in Photobiology" (B. C. Christensen and B. Buchmann, eds.) pp. 422–426, Proc. Third Internat. Congr. Photobiol., Copenhagen, 1960.

Morrow, J. E., 1966, Learning in an invertebrate with two types of negative reinforcement, *Psychon. Sci., 5,* 131.

Morrow, J., and Smithson, B., 1968, Learning in an invertebrate with a positive reinforcement of water, *Psychol. Rep., 22,* 1203.

Morrow, J. E., and Smithson, B. L., 1969, Learning sets in an invertebrate, *Science, 164,* 850–851.

Otsuka, M., Kravitz, E. A., and Potter, D. D., 1967, Physiological and chemical architecture of a lobster ganglion with particular reference to gamma-aminobutyrate and glutamate, *J. Neurophysiol., 30,* 725–752.

Pardi, L., 1960, Innate components in the solar orientation of littoral amphipods, *Cold Spring Harbor Symp. Quant. Biol., 25,* 395–341.

Pardi, L., and Papi, F., 1961, Kinetic and tactic responses, in "The Physiology of Crustacea" (T. H. Waterman, ed.), Vol. II, pp. 365–399, Academic Press, New York.

Pieron, H., 1910, "L'Evolution de la Memoire," Flammarion, Paris.

Pittendrigh, C. S., 1960, Circadian rhythms and the circadian organization of living systems, *Cold Spring Harbor Symp. Quant. Biol., 25,* 159–184.

Reese, E. S., 1963, The behavioral mechanisms underlying shell selection by hermit crabs, *Behaviour, 21,* 78–126.

Remler, M., Selverston, A., and Kennedy, D., 1968, Lateral giant fibers of crayfish: Location of somata by dye injection, *Science, 162,* 281–283.

Roberts, A., 1968, Recurrent inhibition in the giant-fibre system of the crayfish and its effect on the excitability of the escape response, *J. Exptl. Biol., 48,* 545–567.

Sachs, L. B., Klopfer, F. D., and Morrow, J. E., 1965, Reactive inhibition in the sow bug, *Psychol. Rep., 17,* 739–743.

Schmitt, W. L., 1965, "Crustaceans," University of Michigan Press, Ann Arbor.

Schöne, H., 1954, Statocystenfunktion und statische Lageorientierung bei dekapoden Krebsen, *Z. Vergl. Physiol., 36,* 241–260.

Schöne, H., 1961*a*, Complex behavior, *in* "The Physiology of Crustacea (T. H. Waterman, ed.), Vol. II, pp. 465–520, Academic Press, New York.

Schöne, H., 1961*b*, Learning in the spiny lobster *Panulirus argus, Biol. Bull., 121,* 354–365.

Schöne, H., 1964, Release and orientation of behavior and the role of learning as demonstrated in Crustacea, *in* "Learning and Associated Phenomena in Invertebrates" (W. H. Thorpe and D. Davenport, eds.), *Anim. Behav. Suppl., 1,* 135–143.

Schwartz, B., and Safir, S. R., 1915, Habit formation in the fiddler crab, *J. Anim. Behav., 5,* 226–239.

Seabrook, W. D., and Nesbitt, H. H. J., 1966, The morphology and structure of the brain of *Orconectes virilis* (Hagen) (Crustacea, Decapoda), *Can. J. Zool., 44,* 1–21.

Sherman, R. G., and Atwood, H. L., 1971, Synaptic facilitation: Long-term neuromuscular facilitation in crustaceans, *Science, 171,* 1248–1250.

Spaulding, E. G., 1904, An establishment of association in hermit crabs, *Eupagurus longicarpus, J. Comp. Neurol. Psychol., 14,* 49–61.

Stretton, A. O. W., and Kravitz, E. A., 1968, Neuronal geometry: Determination with a technique of intracellular dye injection, *Science, 162,* 132–134.

Takeda, K., and Kennedy, D., 1964, Soma potentials and modes of activation of crayfish motoneurons, *J. Cell. Comp. Physiol., 64,* 165–181.

Takeda, K., and Kennedy, D., 1965, The mechanism of discharge pattern formation in crayfish interneurons, *J. Gen. Physiol., 48,* 435–453.

Taylor, R. C., 1970, Environmental factors which control the sensitivity of a single crayfish interneuron, *Comp. Biochem. Physiol., 33,* 911–921.

Ten Cate-Kazejewa, B., 1934, Quelques observations sur les Bernards l'Ermite (*Pagurus arrosor*), *Arch. Neerl. Physiol. Ser. 3c, 19,* 502–508.

Thompson, R., 1957, Successive reversal of a position habit in an invertebrate, *Science, 126,* 163–164.

Thorpe, W. H., 1956, "Learning and Instinct in Animals," Harvard University Press, Cambridge, Mass.

van der Heyde, A., 1920, Über die Lernfähigheit der Strandkrabbe *Carcinus maenas, Biol. Zentr., 40,* 503–514.

Volz, P., 1938, Studien über das Knallen der Alpheiden, nach Untersuchungen an *Ahpheus dentipes* Guérin und *Synalpheus laeuimanus* (Heller), *Z. Morphol. Ökol. Tiere, 34,*

von Buddenbrock, W., 1953, Nervenphysiologie, *in* "Vergleichende Physiologie," Vol. 2, Birkhauser, Basel.

von Frisch, K., 1953, "The Dancing Bees," Harcourt, Brace, New York.

Warren, J. M., 1965, Primate learning in comparative perspective, in "Behavior of Nonhuman Primates" (A. M. Schrier, H. F. Harlow, and F. Stollnitz, eds.), Vol. I, pp. 249–281, Academic Press, New York.

Watanabe, M., and Iwata, K. S., 1956, Alternative turning response of *Armadillidium vulgare, Ann. Anim. Psychol. Tokyo, 6,* 75–82.

Waterman, T. H., and Chace, F. A., Jr., 1960, General crustacean biology, in "The Physiology of Crustacea" (T. H. Waterman, ed.), Vol. I, pp. 1–33, Academic Press, New York.

Waterman, T. H., Wiersma, C. A. G., and Bush, B. M. H., 1964, Afferent visual responses in the optic nerve of the crab, *Podophthalamus, J. Cell. Comp. Physiol., 63,* 135–155.

Wiersma, C. A. G., 1947, Giant nerve fiber systems of the crayfish. A contribution to comparative physiology of synapse, *J. Neurophysiol., 10,* 23–38.

Wiersma, C. A. G., 1957, On the number of nerve cells in a crustacean central nervous system, *Acta Physiol. Pharm. Neerl., 6,* 135–142.

Wiersma, C. A. G., 1961, Reflexes and the central nervous system, in "The Physiology of Crustacea" (T. H. Waterman, ed.), Vol. II, pp. 241–279, Academic Press, New York.

Wiersma, C. A. G., 1970, Reactivity changes in crustacean neural systems, in "Short-Term Changes in Neural Activity and Behavior" (G. Horn and R. A. Hinde, eds.), pp. 211–236, Cambridge University Press., Cambridge.

Wilson, D. M., and Davis, W. J., 1965, Nerve impulse patterns and reflex control in the motor system of the crayfish claw, *J. Exptl. Biol., 43,* 193–210.

Wilson, D. P., 1949, Notes from the Plymouth Aquarium, *J. Marine Biol. Ass. U.K., 29,* 345–351.

Wine, J. J., 1971, Hyperreflexia in the crayfish abdomen following denervation: Evidence for supersensitivity in an invertebrate central nervous system, Ph. D. dissertation, University of California, Los Angeles.

Wine, J. J., 1973, Invertebrate central neurons: Orthograde degeneration and retrograde changes after axotomy, *Exptl. Neurol.,* in press.

Wine, J. J., and Krasne, F. B., 1969, Independence of inhibition and habituation in the crayfish lateral giant fiber escape reflex, Proc. Seventy-seventh Ann. Convention A.P.A., pp. 237–238.

Wine, J. J., and Krasne, F. B., 1972, The organization of escape behavior in the crayfish, *J. Exptl. Biol., 56,* 1–18.

Woodworth, R. S., and Schlosberg, H., 1956, "Experimental Psychology," Henry Holland, New York.

Wrede, W. L., 1929, Versuche über die Chemoreception bei *Eupagurus bernhardus, Tijdschr. Ned. Dierk. Ver Leiden., 3,* 109–112.

Yerkes, R., 1902, Habit formation in the green crab, *Biol. Bull., 3,* 241–244.

Yerkes, R. M., and Huggins, G. E., 1903, Habit formation in the crawfish, *Cambarus affinis, Harvard Psychol. Stud., 1,* 565–577.

Zucker, R. S., 1972, Crayfish escape behavior and central synapses. I. Neural circuit exciting lateral giant fiber, II. Physiological mechanisms underlying behavioral habituation, III. Electrical junctions and dendritic spikes in fast flexor motoneurons, *J. Neurophysiol., 35,* 599–651.

Zucker, R. S., Kennedy, D., and Selverston, A. I., 1971, Neuronal circuit mediating escape responses in crayfish, *Science, 173,* 645–650.

Chapter 8

LEARNING IN INSECTS EXCEPT APOIDEA

THOMAS M. ALLOWAY[1]

Erindale College
University of Toronto
Mississauga, Ontario, Canada

I. INTRODUCTION

The class Insecta of the phylum Arthropoda contains by far the largest assemblage of organisms in the animal kingdom, approximately 75% of all known living animal species being' insect species (Ross, 1965). When the number of species, the large size of insect populations, and the extremely important effects which these populations have on other organisms are considered together, one might almost go so far as to say that the insects are the dominant life form on this planet. For these reasons, and also because of the importance of insects to agriculture and their role as vectors in the spread of disease, entomology has long been an important subdiscipline of zoology. However, entomological research has largely been directed toward establishing a satisfactory understanding of insect evolution and toward investigating those aspects of insect physiology which are directly related to the control of insect pests. Neither entomologists nor animal behaviorists have given the behavior of this extremely important group of animals the attention it deserves.

Nevertheless, a certain amount of information about insect behavior generally and insect learning in particular has been obtained. In this chapter, a systematic review of laboratory studies of learning in insects other than bees

[1]The author wishes to thank Professors Glenn Morris and David Gibo for their helpful advice and comments.

will be presented, and an attempt will be made to place the findings of these studies in a comparative perspective whenever possible. However, the significance of the learning literature can be better understood after first considering some of the fundamental characteristics of insects, their evolutionary history, and classification.

II. INSECT BODY PLAN

Like all arthropods, insects are basically metameric (segmented) animals having a chitinous exoskeleton which is shed periodically as the animal grows. The principal characteristics which distinguish insects from other arthropods are (1) the division of the body into distinct head, thoracic, and abdominal regions, (2) the presence of three pairs of thoracic legs, and (3) in most orders the presence of wings in the adult (Fig. 1).

The anterior region of the body, the head, is a composite structure which resulted from the evolutionary fusion of several segments. In the typical adult insect, the head bears the compound eyes, the antennae, three pairs of mouthparts (the mandibles, maxillae, and labium), and (when present) the ocelli (Fig. 2). The compound eyes are typically many-faceted organs located on the dorsolateral surface of the head. However, the number of facets varies immensely from species to species, ranging from a single facet in some forms (especially larvae) to over 25,000 facets in some beetles. The ocelli are single-faceted visual receptors which are also present in many adult and most

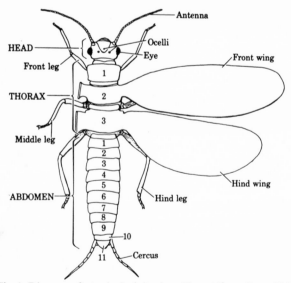

Fig. 1. Diagram of a typical adult winged insect (from Ross, 1965).

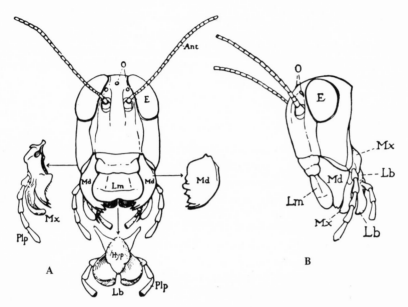

Fig. 2. Diagram of a grasshopper's head. A, Anterior view. B, Lateral view. *Ant*, antenna; *E*, compound eye; *Hyp*, hypopharynx; *Lb*, labium; *Lm*, labrum; *Md*, mandible; *Mx*, maxilla; *O*, ocelli; *Plp*, palpus (from Ross, 1965, after Snodgrass).

larval insects. The antennae are a pair of appendages attached to the front of the head which serve important tactile and chemoreceptive functions in most insects.

The thorax, which constitutes the middle portion of the insect body, always consists of three segments: prothorax, mesothorax, and metathorax. Each thoracic segment bears a pair of locomotor appendages or legs, and except in wingless (apterous) forms, the two pairs of wings are attached to the mesothoracic and metathoracic segments.

The posterior portion of the insect body is the abdomen, a multisegmented structure which is free of locomotor appendages. The basic number of abdominal segments is 11. However, the number of visible segments varies from eight to eleven in different orders (Chapman, 1969).

III. STRUCTURE OF THE INSECT NERVOUS SYSTEM

As in the case in all animals possessing nervous systems, the basic unit of the insect nervous system is the neuron. The typical insect motor or internuncial neuron is monopolar, with its cell body situated in the medullary (peripheral) portion of a ganglion of the central nervous system and an axon extending into the neuropile or central portion of the ganglion (Chapman, 1969; Imms, 1957). However, the peripheral sensory neurons are usually bipolar, with a short dendrite receiving stimuli from the environment and a long axon extending into the neuropile of a central ganglion. The cell bodies of these sensory cells are usually located in or near the sense organs (Chapman, 1969; Imms, 1957). Multipolar neurons also occur in a few regions of the insect central nervous system, and the subepidermal nerve plexus of many larval insects is largely composed of multipolar sensory cells (Chapman, 1969).

As noted above, the cell bodies of motor and internuncial neurons are aggregated in the peripheral portions of the ganglia of the central nervous system (Fig. 3). Many glial cells also have their cell bodies in the peripheral portions of the ganglia. The centers of the ganglia, which are called neuropiles, consist of a complex array of the axons of sensory, motor, and internuncial cells together with their supporting glial processes. The great majority of the synapses in the insect nervous system, and possibly all of them, occur between axons at glia-free sites in the neuropiles of ganglia. Thus the typical insect synapse is axoaxonic in nature. No clear example of the axosomatic synapse, which is common in vertebrates, has yet been found in an insect (Hoyle, 1970; Smith and Treherne, 1963).

The organization of the insect nervous system is fundamentally similar to that of annelids, in that in both phyla the nervous system is basically

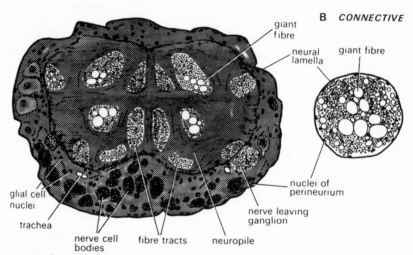

Fig. 3. Diagram of transverse sections of (A) an abdominal ganglion and (B) an abdominal interganglionic connective in the cockroach (*Periplaneta americana*) (from Chapman, 1969, after Roeder).

composed of a series of serially connected segmental ganglia. However, the numerous modifications of the insect nervous system have tended partially to overcome the independence of the certain segments through the fusion of ganglia to form "brains" and "subbrains" (Fox and Fox, 1964).

There are two major neural centers in the head, the supraesophageal ganglion (commonly called the brain) located above the anterior end of the foregut and the subesophageal ganglion located below it (Fig. 4). These two structures are connected by the circumesophageal commissure. The ventral nerve cord, which emerges from the subesophageal ganglion, provides the connection between the anterior centers and the segmental ganglia of the thorax and abdomen.

The number of anatomically separate segmental ganglia varies from one insect order to another and, within some orders, from family to family.

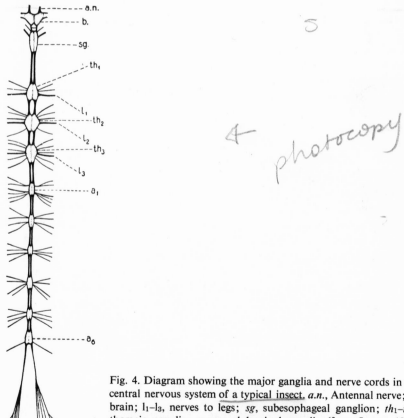

Fig. 4. Diagram showing the major ganglia and nerve cords in the central nervous system of a typical insect. a.n., Antennal nerve; b, brain; l_1–l_3, nerves to legs; sg, subesophageal ganglion; th_1–th_3, thoracic ganglia; a_1–a_6, abdominal ganglia (from Imms, 1959, "General Textbook of Entomology," E. P. Dutton).

An often-repeated evolutionary trend in insects has been for the abdominal and sometimes also the thoracic ganglia to coalesce into a single mass. In a few extreme cases, the thoracic and abdominal ganglia have merged with the subesophageal ganglion. Bullock and Horridge (1965) provide a detailed account of these structural differences in various insect groups.

The supraesophageal ganglion or brain is the most important "association center" in the insect. It receives sensory input from the sense organs of the head and also by means of ascending internuncial fibers from the subesophageal, thoracic, and abdominal ganglia. Motor nerves arising in the brain control the antennal muscles, and premotor internuncial fibers run from the brain to the more posterior centers, thereby controlling the activities of the rest of the nervous system to some extent (Chapman, 1969).

Anatomically, the insect supraesophageal ganglion is divided into three parts, the protocerebrum, the deuterocerebrum, and the tritocerebrum (Fig. 5). The protocerebrum receives input from the eyes and ocelli and contains the complex of neural tissue which seems to be responsible for the processing and analysis of visual sensory data. Also contained within the protocerebrum are the corpora pendunculata, which many authors (e.g., Bullock and Horridge, 1965; Chapman, 1969; Imms, 1957) have viewed as being an important "association area" responsible for mediating complex behavior patterns. The deuterocerebrum contains the antennal lobes, which receive sensory input from the antennae and send out the motor nerves which control the antennal muscles. The tritocerebrum is a pair of small ganglia from which the circumesophageal commissure passes to the subesophageal ganglion. The two halves of the tritocerebrum are connected by the tritocerebral commissure.

The subesophageal ganglion (Fig. 5) is the first ganglion of the ventral nerve chain. It is believed to consist of the fused ganglia of the three hypothetical segments of which the three pairs of mouthparts (mandibles, maxillae, and labium) are the appendages. From it, the nerves which innervate the mouthparts arise and the ventral nerve cord emerges.

The thoracic ganglia give rise to the sensory and motor nerves of the legs and wings. The exact arrangement of these nerves varies considerably with the morphological structures of different insects. Bullock and Horridge (1965) discuss these variations extensively. Each abdominal ganglion innervates the muscles of its own segment.

IV. INSECT EVOLUTION AND CLASSIFICATION

There is general agreement that all arthropods were evolutionarily derived from polychaete annelids. Moreover, it is usually agreed that,

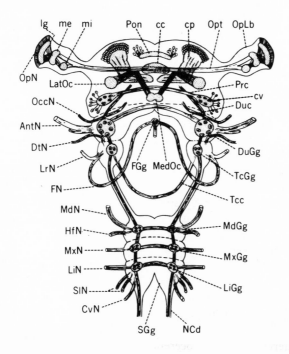

Fig. 5. Schematic diagram of the cephalic part of the nervous system of a typical insect. AntN, antennal nerve; cp, corpus pedunculatum; cc, corpus centrale; cv, corpus ventrale; CvN, cervical nerve; DtN, deuterocerebral tegumental nerve; Duc, deuterocerebral commissure; DuGg, deuterocerebral (antennal) ganglion; FGg, frontal ganglion; FN, frontal nerve; HfN, hypopharyngeal nerve; LatOc, lateral ocellar lobe; lg, lamina ganglionaris; LiGg, labial ganglion; LiN, labial nerve; LrN, labral nerve; MdGg, mandibular ganglion; MdN, mandibular nerve; me, medulla externa; MedOc, median ocellar lobe; mi, medulla interna; MxGg, maxillary ganglion; MxN, maxillary nerve; Ncd, central nerve cord; OccN, occipital nerve; OpLb, optic lobe; OpN, optic nerve; Opt, optic tract; Pon, pons cerebralis; Prc, protocerebral commissure; SGg, subesophageal ganglionic mass; SlN, salivary nerve; Tcc, tritocerebral commissure; TcGg, tritocerebral ganglion (from Fox and Fox, 1964, "Comparative Entomology," Van Nortrand Reinhold).

within the arthropod phylum, rather close affinities exist between the extinct trilobites and modern chelicerates on the one hand and between the modern myriopods and insects on the other. However, owing to the incompleteness of the fossil record, the relationships between the trilobite–chelicerate group and the myriopod–insect group and between these two groups and the crustaceans are unclear. One commonly used system of classification based on the structure of the mouthparts (e.g., Snodgrass, 1938; Fox and Fox, 1964) places the myriopods, insects, and crustaceans together in the subphylum Mandibulata and argues that all modern arthropods are derived from a common pretrilobite arthropod base. However, other systematists (e.g., Manton, 1964; Tiegs and Manton, 1958) argue that the otherwise enormous differences between the crustaceans and the myriopod–insect group can best be accounted for by assuming parallel evolution of the similar mouthparts. For various reasons, adherents to this second view argue for a separate bi-phyletic evolution of two major arthropod groups from the annelid base and

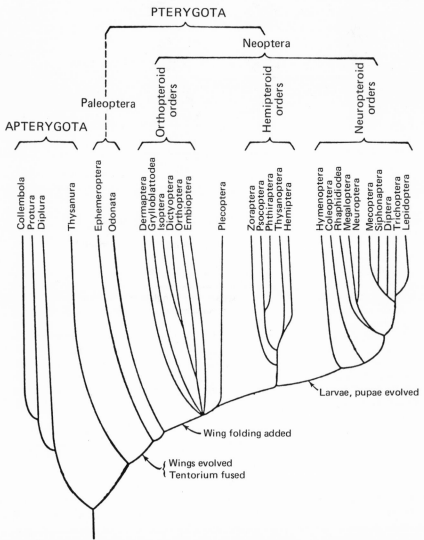

Fig. 6. The insect phylogenetic tree showing the hypothetical relations among the insect orders (modified after Ross, 1965).

believe that the onychophorans (which are sometimes classed as a separate phylum by those who hold other views), myriopods, and insects form one phyletic group, while the trilobites, crustaceans, and chelicerates form another.

However, despite the uncertainty about the unity of the traditional arthropod phylum and also despite a very incomplete fossil record of insect evolution, the survival to the present time of a large number of insect orders which seem to be representative of many stages of insect evolution has permitted the reconstruction of the broad outlines of that evolutionary process

Table I. Outline of Insect Classification (After Ross, 1965)

Subclass Apterygota	Primitively wingless insects
Order Diplura	Small, blind insects found under leaves, stones, or debris
Order Protura	Minute, blind insects which live in the soil or in humus
Order Collembola	Springtails
Order Thysanura	Silverfish and bristletails
Subclass Pterygoga	Winged or secondarily wingless insects
Series Paleoptera	Wings do not fold
Order Ephemeroptera	Mayflies
Order Odonata	Dragonflies and damselflies
Series Neoptera	Wings fold
Order Dictyoptera	Cockroaches, walking sticks, and mantids
Order Isoptera	Termites
Order Orthoptera	Grasshoppers and crickets
Order Dermaptera	Earwigs
Order Grylloblattodea	Small, secondarily wingless insects found at high altitudes on some mountain ranges
Order Embioptera	Web spinners
Order Plecoptera	Stoneflies
Order Zoraptera	Rare, minute insects found under dead bark or rotted wood
Order Psocoptera	Psocids and booklice
Order Phthiraptera	Chewing and biting lice
Order Thysanoptera	Thrips
Order Hemiptera	Bugs
Order Hymenoptera	Bees, wasps, and ants
Order Coleoptera	Beetles
Order Megaloptera	Dobsonflies
Order Neuroptera	Lacewings
Order Rhaphidiodea	Snakeflies
Order Mecoptera	Scorpionflies
Order Siphonaptera	Fleas
Order Diptera	Two-winged (true) flies
Order Trichoptera	Caddisflies
Order Lepidoptera	Moths and butterflies

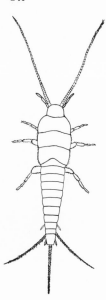

Fig. 7. A member of the primitively wingless order Thysanura, the silverfish *(Lepisma saccharina)* (from Imms, 1959, "General Text-book of Entomology," E. P. Dutton).

from morphological and embryological evidence (supplemented whenever possible by fossil evidence). An approximate outline of the evolutionary relationships among the 28 most generally recognized orders of living insects is presented in Fig. 6, and Table I lists the common names of members of the various insect orders.

The earliest insects are believed to have been wingless creatures which rather closely resembled their centipede ancestors, the most prominent characteristics differentiating them from centipedes being (1) the presence of a three-segmented thorax bearing a pair of locomotor appendages on each of its segments and (2) a great reduction in the number and size of the abdominal appendages (Ross, 1965). In comparison with other insects, the most striking characteristic of such a protoinsect would be the primitive absence of wings. Among the living insects, the Collembola, Protura, Diplura, and Thysanura (Fig. 7) are thought to exemplify this primitively wingless state. Accordingly, they are grouped together in the subclass Apterygota.[2]

An important early advance in insect evolution was the development of wings. The importance of this evolutionary event can scarcely be overesti-mated, since the ability to fly permitted insects to fill a very large number of ecological niches which would otherwise have been closed to them (Fox and Fox, 1964; Ross, 1965). The evolution of wings must have been a prolonged

[2]Other wingless insects (e.g., ants) are thought to be derived from ancestral forms which possessed wings.

process. However, two major stages in this process can be seen in living insects. The first insect wings are believed to have been deeply pleated structures which could not be folded over the body. Wings of this primitive type are found today in members of the orders Ephemeroptera (mayflies) and Odonata (dragonflies and damselflies) (Fig. 8), which are grouped together under the rubric Paleoptera. The foldable wings of most other insects (Neoptera) represent later evolutionary stages.

Another major milestone in insect evolution was the development of holometabolous metamorphosis involving the familiar four-stage (egg, larva, pupa, and adult) life cycle of many advanced insects. This development permitted a single species to become adapted to two different modes of life (a larval mode and an adult mode) and thus to exploit seasonal and cyclic changes in such critical factors as the food supply. Holometabolous metamorphosis and flight together account for much of the great evolutionary radiation and proliferation of insect forms.

As we shall see, laboratory studies of learning have been confined to species belonging to only a few of the neopteran orders. No studies of species belonging to the paleopteran or apterygotan orders have been performed. Of the groups which have been studied, the ants and wasps (Hymenoptera), fruit flies (Fig. 9) and blowflies (Diptera), and grain beetles (Coleoptera) (Fig. 10) are holometabolous. On the other hand, the cockroaches (Dictyoptera) (Fig. 11) and locusts (Orthoptera) show what is known as hemimetabolous metamorphosis. Hemimetabolous insects do not show clearly distinguishable larval and pupal stages. In contrast, the immature hemimetabolous insect resembles the adult insect rather closely, the chief differ-

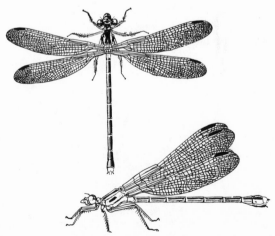

Fig. 8. A member of the paleopteran order Odonata, the damselfly *(Archilestes californica)* (from Ross, 1965).

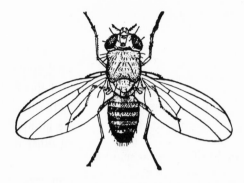

Fig. 9. An adult fruit fly (*Drosophila melanogaster*) (from Ross, 1965).

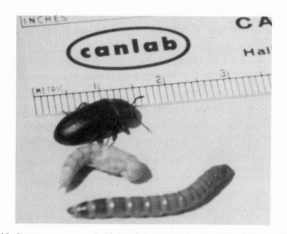

Fig. 10. Larva, pupa, and adult of the grain beetle (*Tenebrio molitor*).

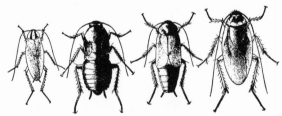

Fig. 11. The three species of cockroach which have been most frequently used in laboratory studies of learning: *Blattella germanica, Blatta orientalis* male and female, and *Periplaneta americana* (from Ross, 1965).

ences (other than size) between immature (nymph) and adult being the absence of completely developed wings and sexual organs in the immature.

Finally, the evolution of a social mode of life has occurred in a number of different insects, primarily in the orders Isoptera and Hymenoptera. Space permits only the briefest mention of insect social life in this chapter. However, two points should be made. First, unlike the other evolutionary advances which have been described so far, social life is believed to have been attained through parallel evolution in the several different families in which it occurs (Wheeler, 1928; Lanham, 1964). Second, social life represents a rather extreme form of specialization not characteristic of the vast majority of insects. Thus it is among the vast hordes of nonsocial insects where "typical" insect behavior is to be found.

V. INSECTS AS SUBJECTS FOR LABORATORY INVESTIGATIONS OF BEHAVIOR AND NEUROPHYSIOLOGY

The reasons for choosing an insect as the subject of a behavioral or neurophysiological investigation vary with the aims of the investigator. On the one hand, as noted at the beginning of this chapter, the insects are the nost numerous and in some senses the dominant group of metazoan animals. Moreover, their activities are of major practical importance to man. These facts mean that the study of insect behavior and of its neural and other correlates is justifiable not only as "pure science" but also by virtue of its possible useful applications in such fields as human and veterinary medicine, agriculture, and forestry. In other words, behavioral entomology and insect neurophysiology are justifiable enterprises in their own right because of the knowledge they provide about an extremely important group of animals. For this reason, some investigators derive an interest in insect behavior and neurophysiology from a broader interest in entomology, and for them the question of why insect behavior should be studied has an obvious answer.

However, perhaps more commonly, investigators of insect behavior and neurophysiology owe allegiance to comparative psychology, to ethology, or to that subbranch of neurophysiology which seeks to find general principles of neural organization. Investigators of this second type are commonly interested in studying some process (e.g., habituation or classical conditioning), and they tend to look for animals which will serve as convenient and representative preparations for the study of that process. For them, the insects, like most other animal groups, offer a number of advantages as well as certain disadvantages. The discussion which follows will consider primarily those characteristics of insects which make them particularly useful and particularly difficult subjects for the study of learning and its neural correlates.

One of the main obstacles which scientists interested in the study of the neurophysiology of learning have encountered is the enormous complexity of the vertebrate nervous system. As an essential prerequisite for the investigation of the cellular correlates of learning, one needs to localize the physiological processes producing the behavioral change in a mass of tissue small enough for practical analysis. The problem is rather like the proverbial task of locating a needle in a haystack, and vertebrate nervous systems with their 10^{10} or more cells are very large haystacks indeed.

Insects, like other invertebrates, are attractive to scientists interested in the neurophysiology of learning because they possess many fewer neurons and thus constitute a much smaller haystack. The typical insect nervous system has only about 10^5 to 10^6 neurons. Moreover, since perhaps 80% of these neurons seem to be involved with the analysis of sensory input (Hoyle, 1970), the number of neurons responsible for the integration of motor output is relatively small, yet it is among this relatively small number of cells where one would expect most changes associated with learning to be located.

In addition, the segmental organization of much of the insect nervous system offers the prospect of localization of at least some forms of learning in a single segmental ganglion. As we shall see, this prospect appears to have been realized in at least one important instance (Eisenstein and Cohen, 1965; Hoyle, 1965).

Insect metamorphosis offers another possible approach to the localization of cellular changes associated with learning and memory storage. In holometabolous insects, many, but not all, of the larval neurons are replaced by new adult neurons during the pupal stage of life, the new neurons being derived from persistent postembryonic neuroblasts (Bullock and Horridge, 1965; Edwards, 1969, 1970). Nevertheless, retention of learning through metamorphosis occurs in at least some insects (Alloway, 1972; Borsellino et al., 1967, 1970; Cherkashin et al., 1968), and it would appear likely that the memory for the learned behaviors is stored among the neurons which persist through metamorphosis.

However, despite these advantages, insects offer certain disadvantages for those who seek neural correlates of learning. Perhaps the most significant of these is the small size of insect neurons, a feature which prevents intracellular recording of neural activity under any but the most favorable conditions (Hoyle, 1970; Kandel et al., 1970). The insect's tough chitinous exoskeleton also presents a formidable barrier which must be penetrated before the electrical activity of the nervous system can be recorded.

For those investigators who are primarily interested in performing behavioral analyses of the learning capacities of intact and unrestrained animals, the typical insect's capacity for flight poses another serious problem. This problem is particularly serious in the study of instrumental conditioning, where the maze is one of the standard pieces of laboratory apparatus. It is no accident that most insect species used in laboratory studies of learning either are apterous or have the capacity for flight poorly developed.

Bearing these preliminary considerations about the basic characteristics of insects in mind, we are now ready to begin to consider the insect learning literature.

VI. HABITUATION

Although the number of studies performed is not large, habituation has been observed in insects both at the behavioral level and at various neural levels. Baxter (1957) demonstrated habituation of the running response elicited by puffs of air applied to the anal cerci of the cockroach *(Periplaneta americana)*, and Hollis (1963) demonstrated habituation and dishabituation of the abdominal reflex in pupae of the grain beetle *(Tenebrio molitor)*. This latter study is of particular interest because it represents what may be the only demonstration of behavior modification in a pupal insect. The abdominal reflex in *Tenebrio* pupae consists of a twitching of the entire abdomen which can be elicited by a wide variety of stimuli such as electric shock, tactile stimulation, or onset of a bright light. In this study, the stimulus to which the animals were habituated was a 0.016-ma electric shock administered through wires pressed to the sides of the head. Shock duration was $\frac{1}{2}$ sec, with an interstimulus interval of 30 sec. Habituation was carried to the criterion of ten successive failures to respond to the shock, the pupae requiring a mean of 38.1 shocks to reach this criterion. Then dishabituation was attempted by touching the animal with a brush 1 sec prior to shocking it. The tactile stimulus raised the probability of the shock's eliciting a response to 0.50.

Habituation of the neural precursors of the running response elicited by stimulation of the anal cerci of cockroaches *(Periplaneta americana)* has also been investigated. The ventral nerve cords of cockroaches, certain Orthoptera, and various other insects contain a small number of axons which are very much larger than the other axons and which accordingly are called giant fibers (Fig. 3). In the cockroach, the largest of these giant axons are postsensory internuncial fibers which arise in the last abdominal segment and terminate in the thoracic ganglia. These axons receive stimuli from the sensory neurons of the cercal sensillae. In the abdominal ganglia, the giant fibers synapse with other internuncial cells which are also involved in mediating the running response (Chapman, 1969; Hoyle, 1970; Rowell, 1970).

The rest of the neural circuitry involved in mediating the running response has not yet been worked out in detail; however, at least four or five synapses are probably involved (Hoyle, 1970).

Pumphrey and Rawdon-Smith (1936, 1937) measured the neural activity produced in the metathoracic ganglion by repetitive stimulation of the anal cerci and observed that what they called "adaptation" occurred. In a later study, Niklaus (1965) measured the activity of the cercal sensory cells and of the giant fibers during continuous and repetitious stimulation of the cerci. Although even continuous stimulation produced no response decrement in the sensory cells, the activity of the giant fibers declined rapidly. Similar observations were made with repetitive stimuli having interstimulus intervals of less than 5 min. This decrement in giant fiber activity may account for some of the behavioral habituation of the running response. However, other evidence suggests that some of the other internuncial neurons may also be involved. For example, Hoyle (1970) reports an unpublished study performed in his laboratory in which the giant fibers were stimulated electrically. Under these circumstances, habituation of the muscle potentials associated with the running response still occurred even though there was no decrement in giant fiber activity.

A number of other investigators (e.g., Horn and Rowell, 1968; Horridge *et al.,* 1965; Rowell and Horn, 1968) have reported habituation of neural responses evoked by visual stimulation of the compound eyes in various parts of the locust *(Schistocerca gregaria)* supraesophageal ganglion. Horn and Rowell (1968) recorded the activity of visual internuncial neurons in the locust tritocerebrum. Most of these units responded to objects moving in the contralateral visual field. Response decrements were observed with interstimulus intervals as long as 10 min. Dishabituation of the activity of these same units was produced by electrical stimuli which did not elicit a response in the habituated units (Rowell and Horn, 1968).

VII. CLASSICAL CONDITIONING

Studies claiming or appearing to be straightforward applications of the classical conditioning paradigm with insects as subjects have been rare, although several previous reviewers (e.g., Schneirla, 1953; Thorpe, 1943; Warden *et al.,* 1941) have invoked classical conditioning as an explanatory mechanism to account for certain phenomena of insect behavior. Moreover, most of the studies which have appeared have employed bees as subjects and will thus be discussed in the next chapter. Indeed, the only two experiments known to me which might be viewed as reasonably straightforward applications of classical conditioning procedures to insects other than *Apis mellifera* are those by Frings (1941) which the blowfly *(Cynomyia cadaverina)* and Nelson (1971) with the blowfly *(Phormia regina)*.

Frings (1941) used the odor of cumarin as a CS which was paired with immersion of the feet in a sucrose solution. The latter stimulus is an effective UCS for eliciting proboscis extension in *Cynomyia cadaverina*. In the absence of experience with pairing the odor of cumarin with the presentation of sucrose, cumarin elicited proboscis extension on 4% of trials. Training consisted of alternating test trials during which only cumarin was presented with training trials during which cumarin was paired with sucrose. During training, cumarin elicited proboscis extension on 73% of the first ten test trials and on 90% of the last ten test trials after 64 CS–UCS pairings. However, since Frings was interested only in determining the location of his subjects' olfactory receptors and not in studying their learning capacities as such, no controls for sensitization or pseudoconditioning were included. However, the absence of such controls renders his findings inadequate as a demonstration of classical conditioning. In fact, the discovery of Dethier and his associates (Dethier, 1969; Dethier *et al.,* 1965) that stimulation with sucrose causes the blowfly to show proboscis extension to stimuli which would not otherwise elicit that response made it appear very likely that Frings' (1941) findings could be attributed to pseudoconditioning.

However, the series of experiments described by Nelson (1971) offer much more convincing evidence for conditioning in the blowfly *(Phormia regina)* inasmuch as an explicit effort was made to control for both sensitization and pseudoconditioning. In these studies, the UCS which elicited proboscis extension was application of a small droplet of sucrose solution to the mouthparts, and water or saline solution applied to the feet was used as CSs. Immediately before training, the animals were satiated for water so that the CSs only very rarely elicited proboscis extension as an unconditioned response.

In her first experiment, Nelson (1971) studied sequential compound conditioning. One (experimental) group of 200 flies received both CSs one after the other for 4 sec each, with the UCS being applied 1 sec before termination of the second CS. Half the subjects received the CSs in the order water–saline, while the other half received them in the reverse order. A second group of 200 flies served in a sensitization and thirst control condition. These animals were treated exactly like those in the first group except that the UCS was not presented. All animals received 15 presentations of the two CSs. The results for the animals in the first group showed an increasing tendency for both CSs to elicit proboscis extension as a function of trials, with responsiveness to the second CS, which more closely preceded onset of the UCS, being consistently greater than responsiveness to the first CS. There was no general tendency toward increasing responsiveness to either CS in the sensitization and thirst control group. The greater responsiveness of animals in the first group to the second CS strongly, albeit indirectly, suggests that these results are not attributable to pseudoconditioning. If the increasing responsiveness of these animals had not been dependent on the pairing of the CSs with UCS, then one would have expected either that the animals would be equally responsive to the two CSs or that they would be more responsive to the first CS owing to the greater temporal proximity of the first CS to termination of the UCS on the previous trial.

Moreover, Nelson (1971) reported six other experiments which offered further evidence of classical conditioning in *Phormia regina*. Among other things, these studies included demonstrations that no progressive or sustained increase in responsiveness to the CSs results either from repeatedly presenting the UCS alone or from presenting all three stimuli in an unpredictable and unpaired sequence. These demonstrations make it appear even more unlikely that the results of the study described above could have been due to either pseudoconditioning or sensitization. A demonstration of stimulus discrimination adds further weight to this conclusion.

VIII. INSTRUMENTAL CONDITIONING: TASKS AND SPECIES

A. Cockroaches

Two situations suitable for the demonstration of instrumental conditioning in cockroaches have been developed. The first of these was devised independently by Szymanski (1912) and Turner (1912) in the same year. The apparatus consists of closed box mounted under an overhead light. Half the box is transparent, and half is opaque to produce relative darkness. When a roach is introduced into this box, it seeks shelter from the light in the darkened compartment. However, the floor of the dark area is an electric grid through which shocks are administered, causing the animal to run back into the lighted area. After varying amounts of experience with the shock, the animals learn to remain in the lighted area. Various versions of this procedure have been used to train specimens of *Blatta orientalis*[3] (Szymanski, 1912; Turner, 1912), *Periplaneta americana* (Freckleton and Wahlsten, 1968; Minami and Dallenbach, 1946), and *Blattella germanica* (Hunter, 1932). Reports of the duration of memory vary considerably, but the learning probably persists for at least 48 hr (Minami and Dallenbach, 1946) and perhaps for as long as 9 days in some cases (Szymanski, 1912). At least one group of investigators (Ebeling *et al.*, 1966) believes that an avoidance

[3]The reviewer believes that the species which Szymanski (1912) and Turner (1912, 1913) called *Periplaneta orientalis* was in fact *Blatta orientalis*.

learning process analogous to that demonstrated by this procedure is responsible for the learned avoidance of some blatticides which cockroaches sometimes exhibit.

Razran (1971) has concluded that the form of learning observed in this situation is an example of aversive inhibitory conditioning in which one response (the negative phototropism in this case) is inhibited by another antagonistic response (the response to shock) but without any new behavior being learned. However, I disagree with Razran, because new behaviors do in fact seem to be learned. Both Szymanski (1912) and Turner (1912) describe behavior changes which appear to amount to more than the mere cessation of the roach's unconditioned tendency to run to the dark half of the apparatus, and I have confirmed the accuracy of their descriptions in my own laboratory (Alloway, unpublished observations, 1971).

When a roach is first placed in the lighted half of the box, it runs to the darkened half almost immediately. The animal's first response to the shock is *not* to run back into the illuminated area. Instead, it scurries about with great agitation in the dark area and attempts to climb the walls. Escape from the first shock occurs only after some time (up to 5 min). On subsequent trials, the agitated scurrying and attempted wall climbing cease, and the roach eventually runs back toward the lighted area almost immediately. Thus a phase of instrumental escape learning appears to occur.

At the same time, other new behaviors are developing. The animal begins to pause and palpate with its antennae before entering the dark half of the box. Later, it begins to turn away from the darkened area. And eventually it will struggle to keep away from the darkened area if the experimenter pushes it in that direction. With the possible exception of the last example, none of these new behaviors resemble the unconditioned reaction to the shock.

Thus Razran's (1971) conclusion that the behavior change seen amounts to nothing more than inhibition of the negative phototropism by the response to the shock appears to be overly simplistic, and I am inclined to think that genuine avoidance learning occurs. Turner's (1912) demonstration that the learned tendency to remain in a lighted area is limited to the experimental situation and thus is under the control of environmental stimuli is also very much in harmony with an avoidance learning interpretation.

A number of maze situations have also been employed for training cockroaches. The most widely used technique was first employed by Turner (1913). It consists of a flat, elevated, metal maze mounted over water. An overhead light serves as a negative reinforcer from which the animal can escape by entering its home cup, which is mounted at the goal end of the maze. Escape from the situation altogether is prevented by the water into which the roach falls if it leaves the runways of the maze. In this situation, learning, as evidenced by a decrease in time required to traverse the maze or (more commonly) by a decrease in errors as a function of trials, has been observed in *Blatta orientalis* (Turner, 1913), *Blattella germanica* (Chauvin, 1947; Goustard, 1948; Hullo, 1948; Le Bigot, 1952, 1954; Lecompte, 1948), and *Periplaneta americana* (Gates and Allee, 1933). Specimens of *Periplaneta americana* have also been successfully trained to escape from light in a slightly different nonelevated maze situation (Eldering, 1919), and Longo (1964) trained *Nauphoeta cincera* in a simple Y-maze in which escape from electric shock and shock avoidance were the reinforcers. Finally, Longo (1970) reported successful training of *Byrsotria fumigata* in a straight alley using food (crushed pineapple) as a reinforcer. However, two earlier attempts to train *Periplaneta americana* using food reinforcement yielded inconclusive findings in one case (Eldering, 1919) and clearly negative results in the other (Gates and Allee, 1933). In both of the latter cases, the nature of the food used was not specified.

Two important motivational factors in the elevated maze situation are the light from which the roach escapes and the cup into which it escapes. Learning efficiency is impaired if the light is either too dim or too bright (Chauvin, 1947), and roaches will not learn to go to a transparent cup (Goustard, 1948). The construction of the cup and the odors it contains are also important. Roaches prefer cups made of porous materials which are poor conductors of heat (Lecompte, 1948), and they will not learn to go to a cup unless it contains roach odor (Chauvin, 1947; Lecompte, 1948). Further evidence of the attractiveness of a home cup scented with roach odor is the fact that roaches will learn a maze in the absense of light where the only apparent reinforcement is entry into a cup containing roach odor (Lecompte, 1948). The addition of electric shock as punishment for entry into blind alleys has complex and contradictory effects depending on whether the shock is administered in all blind alleys or only in some (Chauvin, 1947; Le Bigot, 1954).

So far as conditions of the task are concerned, there is no simple relationship between the number of blind alleys in a maze and whether or not the maze can be learned. Chauvin (1947) concluded that difficulty in learning multiple T-mazes varied as a cubic function of the number of T units in the maze pattern. However, this relationship could hardly be expected to hold for all possible maze patterns. Hullo's (1948) discovery that Schneirla's (1929) concept of centrifugal swing[4] can successfully account for differences in difficulty in eliminating errors at particular choice points would appear to offer a more fruitful approach to the problem of maze difficulty. Finally,

[4]See the section on ant learning for a more complete discussion of centrifugal swing.

at least when escape from light is used as the reinforcer, speed and degree of learning are enhanced by increasing distribution of practice within and between training sessions (Chauvin, 1947; Le Bigot, 1952). Many details of the distribution effect remain to be worked out, but increasing intertrial intervals up to 20 min within training sessions and increasing intervals between sessions up to 2 hr have been found to be beneficial (Le Bigot, 1952).

Evidence concerning the sensory cues employed by cockroaches in learning mazes is inconclusive. Learning apparently does not occur unless at least one antenna is present and free to explore the environment (Chauvin, 1947; Hullo, 1948). Vision also seems to play some role, since rotating the maze in a way which changes the directionality of illumination interferes with learning (Chauvin, 1947) and with performance of a well-learned maze habit (Eldering, 1919). However, vision may not be an essential factor, since covering the eyes with varnish prior to training does not impede learning (Chauvin, 1947). Chemical trails do not appear to be an important factor. Roaches learn no better in uncleaned mazes then in mazes cleaned with alcohol before each trial (Chauvin, 1947; Eldering, 1919).

Finally, Gates and Allee (1933), in a frequently cited study, claim to have shown that cockroaches (*Periplaneta americana*) learn a maze better when running alone than when running in groups of two or three. However, the results of this study cannot be taken seriously for two reasons. First, all the animals in the study ran the maze alone initially, then ran with one other subject, and then ran with two other subjects. Since social grouping was thus confounded with other actual or potential independent variables such as age of the subjects, transfer of learning, and seasonal changes which might affect behavior, one cannot tell whether the behavioral differences reported should be attributed to social grouping or to one of these other variables. Second, the light from which the subjects escaped was red and may thus have been essentially invisible to them. Perhaps because of this factor, the level of learning attained was not very high in any of the conditions.

Zajonc et al. (1969) have recently reported results with *Blatta orientalis* which partially confirm and extend the Gates and Allee (1933) findings. These investigators measured the running speeds of different groups of roaches in cross-mazes and straight alleys when the animals were running alone, running with one other roach, or running alone but with an audience of other roaches. Both kinds of social grouping were found to inhibit performance in the cross-maze but to facilitate performance in the straight alley. However, Zajonc et al. (1969) failed to present their data in a way which permits any determination of the manner in which their independent variables affected learning.

B. Grain Beetles

Four slightly different training techniques have been used to train both adults and larvae of the grain beetle *(Tenebrio molitor)* in simple two-choice maze situations. Alloway (1969, 1970; Alloway and Routtenberg, 1967) and Walrath (1970) have produced simple left *vs.* right position discriminations using escape from light into a home cup of cereal as a reinforcer. Similar position discriminations have been demonstrated by Borsellino *et al.* (1967, 1970) and by Cherkashin *et al.* (1968) using avoidance of light and electric shock punishment, respectively, as reinforcers. And von Borell du Vernay (1942) produced a textural discrimination using electric shock punishment. In all cases, retention of the position discriminations persisted for periods varying from several days to several weeks. Moreover, three studies (Alloway, 1972; Borsellino *et al.,* 1967, 1970; Cherkashin *et al.,* 1968) have shown retention of memory through metamorphosis; that is, adult beetles retained a maze habit which they had learned as larvae. As noted previously, since most larval neurons are replaced during metamorphosis, retention of memory through metamorphosis may offer a new approach to locating sites in the insect where memory storage occurs.

C. Fruit Flies

Murphy (1967, 1969) has developed an interesting technique for training fruit flies *(Drosophila melanogaster)* to turn left or right consistently in a T-maze. His subjects were members of a negatively geotropic strain, and the reinforcement used was the opportunity to climb a vertical tube located at the end of the correct goal arm of the maze. Learning occurred with this positive reinforcer as the only incentive; however, learning improved as the intensity of an electric shock punishment for incorrect turning increased up to the point where electrocution resulted (Murphy, 1969). Varying the length of the vertical tube in which the animals climbed and the length of the goal arms of the maze had no effect on learning (Murphy, 1967).

D. Ants

The most commonly used training procedure for ants has been a complex maze situation originally developed by Shepard (1911) and modified by Schneirla and his associates (Schneirla, 1929, 1933 1934, 1941, 1943, 1960, 1962; Weiss and Schneirla, 1967) and by others (e.g., Evans, 1932) in an extensive series of investigations of learning and orientation in various species of the *Formica* genus. In these studies, the animals lived in an artificial nest connected to the maze by runways and gates which enabled the experimenter to control access to the experimental situation. Opposite the nest was a food source which the ants could reach by running either through the maze or, in certain experiments, by means of a direct route. Return to the nest was also controlled and could be directed through the maze or over the direct route. Artificial lighting permitted the control of external visual cues, and chemical cues were controlled by interchanging or replacing bristol board alley linings in studies in which the use of chemical cues was to be excluded. Individually marked subjects were then permitted to pass to and from the food source, and learning was measured as a reduction in the number of errors made in running the maze.

In this maze situation, ants apparently make use of visual, chemical, and kinesthetic cues all at the same time, provided all are available. Evidence for the use of visual cues is provided by the fact that changing the direction of directional illumination disrupts performance of a well-learned maze habit (Schneirla, 1929). Chemical cues appear to be somewhat less important. Exchange of alley linings to make chemical trails a useless tool disrupts maze running in the early stages of learning (Schneirla, 1929). Having chemical cues present and intact apparently serves mainly to reduce circuitous running about and repetitive retracing on the first few trials (Schneirla, 1941). The elimination of specific blind alleys does not seem to be facilitated by chemical cues (Schneirla, 1941). The importance of kinesthetic cues is shown by the fact that when a section of the true path of a well-learned maze is lengthened, the ants attempt to make turns at the point where a choice point was previously located (Schneirla, 1929).

In a complex maze, the difficulty of any particular choice point between a blind alley and the true path is profoundly influenced by a factor which Schneirla (1929) called centrifugal swing. This refers to the fact that if a rapidly running ant makes two consecutive turns in the same direction, momentum will throw it into close contact with the outside wall of the last alley in the series and favor a turn to that side at the next choice point. Thus if a U-shaped approach to a T-form choice point forces an ant to make two successive right turns, the animal will be strongly disposed to turn left at the choice point. Conversely, if the animal were forced to make two successive left turns, it would be equally strongly disposed to turn right. Elimination of blind alleys, entry into which is favored by centrifugal swing, is usually a prolonged process. If the blind alley has two arms in the shape of an L, entrance into the arm of the blind alley farthest from the choice point ceases after the first few trips. However, elimination of entry into the arm nearest the choice point proceeds slowly, with the animal making progressively shorter forays into the blind alley until finally it does not enter at all (Schneirla, 1929).

Hoagland (1931) and Vowles (1964, 1965) used much less complex training situations with ants. Hoagland (1931) trained specimens of *Componotus herculeanus pennsylvanicus* de Geer to escape the odor of peppermint in a small shuttle box in which the correct response simply consisted of running from one compartment to the other. Vowles (1964,1965) trained *Formica rufa* in a T-maze using escape from peppermint and return to the home nest as reinforcers. In these studies, Vowles demonstrated that ants can learn a visual discrimination between horizontal and vertical stripes (Vowles, 1965) and olfactory discriminations between various volatile substances (Vowles, 1964). Finally, van der Heyde (1920) trained ants of an unspecified species to jump from a vinegar-saturated platform to a vinegar-free substratum. That the animals were jumping (not falling) from the platform was demonstrated by the fact that they would not acquire the response if the substratum was also saturated with vinegar.

E. Wasps

Rabaud (1926) and Verlaine (1924, 1925, 1927) are members of a group of investigators who have placed colonies of wasps in various artificial situations and observed how the workers learned to make various sensory discriminations and to traverse various apertures and tubes in order to make foraging flights to and from their nests. Generally, these studies have been less well controlled than those which have been reviewed so far, and in many cases the investigations border on naturalistic field observations. Thus they will not be reviewed in detail. However, it should be mentioned that it would appear that wasps are capable of instrumental learning and that they can use information derived from relatively complex visual stimuli and at least simple olfactory stimuli to direct their learned behavior.

IX. "SHOCK AVOIDANCE LEARNING" IN CERTAIN INSECT PREPARATIONS

Horridge (1962*a,b*) developed a technique for training leg flexion in decapitated cockroaches *(Periplaneta americana)* and locusts *(Locusta migratoria)*. Pairs of headless animals were suspended over containers of electrified saline solution. One animal in each pair was shocked whenever its leg fell below a predetermined level defined by the surface of the saline solution. Since in this animal shock administration was contingent on leg *position,* animals serving in this condition have come to be known as "P" animals. The other animal in each pair served as a yoked control, shocks being administered simultaneously with shocks to the "P" animal and without regard to the position of the second animal's leg. Since shocks were delivered more or less at *random* with respect to leg position, the yoked control animals have come to be referred to as "R" animals. After "P" and "R" training had continued in this fashion for 30–45 min, the two animals were unyoked, and each was shocked independently according to the procedure followed for the "P" animal in the first phase of the experiment. During this second testing phase of the study, the "P" animals received significantly fewer shocks than the "R" animals. Inasmuch as the "P" and "R" animals had received the same pattern and intensity of shocks in the initial phase of the experiment, the subsequent difference in their performance was attributed to learning of the contingency between leg position and shock in the "P" animals. Since decapitation had removed the supra- and subesophageal ganglia (brain), leaving only the ventral nerve cord and its associated ganglia intact, it was concluded that the ventral nerve cord had been shown to be capable of mediating a form of learning.

These findings have been confirmed using different response measures in the same preparation of *Periplaneta americana* (Eisenstein, 1970a) as well as in animals whose central nervous system is intact (Disterhoft *et al.,* 1968; Pritchatt, 1968). Moreover, intact animals have been trained to extend their legs as well as flex them (Pritchatt, 1968). In the headless preparation, retention has been demonstrated for as long as 12 hr in a published experiment (Eisenstein, 1970b), and Eisenstein (1971) cites unpublished data from several laboratories indicating retention up to 72 hr. In the intact preparation, retention up to 24 hr has been reported (Disterhoft *et al.,* 1968).

However, of particularly great importance to scientists interested in the molecular bases of learning and memory was the discovery by Eisenstein and Cohen (1965) that learned leg flexion could be mediated by a single ganglion in the ventral nerve cord of *Periplaneta americana.* These investigators applied Horridge's (1962a,b) "P" and "R" training and testing procedures to train prothoracic leg flexion in animals whose prothoracic ganglion had been isolated by decapitation and severance of posterior connections. The conclusion that learning occurred in this preparation was based on three observations: (1) the "P" animals took significantly fewer shocks than the "R" animals during the testing phase of the experiment, (2) the "P" animals took significantly fewer shocks during the test than during the equivalent period of original training, and (3) acquisition of the leg flexion response by the "R" animals during the test was significantly inferior to that of the "P" animals during original training. This last finding was interpreted as indicating that the "R" animals had learned that shock administration was uncorrelated with leg position and indicating that this learning interfered with acquisition of the flexion response during the test. The contention that the learning observed was being mediated by the prothoracic ganglion was strengthened by

the demonstration that there were no significant differences in the performance of "P" and "R" animals from which the ganglion had been removed
prior to training.

At about the same time that Eisenstein and Cohen (1965) reported their
findings, Hoyle (1965) presented an important analysis of the electrophysiological correlates of learning in the thoracic ganglion of cockroaches
(*Periplaneta americana*) and locusts (*Melanoplus differentialis* and *Schistocerca gregaria*) and demonstrated an electrophysiological analogue to the
behavioral learning observed by Eisenstein and Cohen (1965). First, Hoyle
(1965) showed that the conditioned flexion response is usually accompanied
by a prolonged electrical discharge in the coxal levator muscles. Then, while
recording from a motoneuron supplying the levator muscles, he delivered
leg shock whenever the firing rate of the motoneuron fell below a certain
level. In this way, it was possible to induce the neuron to maintain a high
rate of firing. Random or regular shocks not contingent on firing rates did
not lead to a sustained increase in firing rates. Moreover, similar evidence of
learning was obtained in preparations of *Melanoplus differentialis* in which
the appropriate muscles had been dissected away from all connection with
the campiform sensillae and hair plates, which are thought to mediate proprioception of leg position. Thus, at least in this locust and by inference
probably also in the cockroach, proprioceptive feedback appears not to be
an important factor in mediating the learning. This last point is important,
as will be demonstrated in the analysis of the headless insect learning paradigm which follows this discussion.

Recently, Aranda and Luco (1969) have reported results which extend
the findings of Eisenstein and Cohen (1965) and Hoyle (1965). These investigators report that leg flexion learning can be mediated by the metathoracic
ganglion of *Blatta orientalis*. After training, the metathoracic ganglia of
some animals were excised and maintained *in vitro*. Recordings from the
trained side of ganglia from "P" animals revealed steady discharges of 15–
30 spikes per second in the fifth metathoracic nerve. No comparable discharges were recorded from nerves on the untrained side of ganglia from "P"
animals or from any of the nerves of ganglia from "R" animals. However,
owing to the small number of animals studied, these results should be viewed
as tentative.

The effects of various pharmacological agents on the learned leg flexion
behavior of the isolated thoracic ganglion preparation have been the subject
of several investigations. However, the results of these studies are either
extremely tentative (Eisenstein, 1965) or ambiguous owing to the absence of
necessary behavioral control conditions (Brown and Noble, 1967, 1968;
Glassman *et al.*, 1970). Eisenstein (1968) has provided methodological sug-
gestions which should be useful to future investigators in this area.

Recently, Kerkut *et al.* (1970) have reported results concerning chemical changes in the metathoracic ganglion of *Periplaneta americana* during acquisition and retention of leg flexion learning. After the metathoracic legs of "P" and "R" animals had been trained, their metathoracic ganglia were excised and subjected to analysis to determine activity levels for turnover of ribonucleic acid (RNA), protein synthesis, cholinesterase activity, and production of γ-aminobutyric acid (GABA). The ganglia of "P" animals showed increased RNA turnover and protein synthesis as compared to the ganglia of "R" animals and untrained controls. However, cholinesterase activity and production of GABA were lower in the ganglia of "P" animals than in those of "R" animals, and lower in the ganglia of "R" animals than in those of untrained controls. Furthermore, the decrement in cholinesterase activity diminished and returned to normal over a 3-day period, the recovery closely paralleling forgetting of the learned response as evidenced by reduced savings in subjects whose ganglia were not excised. In the light of other evidence identifying actylcholine and GABA, respectively, as probable excitatory and inhibitory transmitter substances in the cockroach central nervous system, Kerkut *et al.* viewed the decrease in cholinesterase and GABA activity levels as probable indication of facilitation of synaptic transmission in some of the neurons of the metathoracic ganglion. They concluded: "Our demonstration of a greater turnover of RNA, enhanced protein synthesis, a rapid decrease in cholinesterase activity and a slower production of GABA in experimental animals leads us to suggest that the latter two changes constitute the basis of short-term memory (1–3 days). Moreover, the changes in protein synthesis may bring about morphological changes that would be responsible for longer term memory" (p. 723). Unfortunately, assessment of the significance of these extremely provocative findings is hampered somewhat by a failure to indicate whether or not the metathoracic ganglia had been isolated prior to training.

Despite this difficulty, results such as those of Kerkut *et al.* (1970) make it appear likely that the molecular processes underlying this form of learning may soon be discovered. If this proves to be the case, questions will arise concerning the plausibility of assuming that the processes mediating learning and memory in headless insects and other invertebrates are similar to processes underlying learning and memory in vertebrates. Ultimately, of course, questions about the similarity of these mechanisms will be answered empirically. However, owing to the complexities of vertebrate nervous systems, definitive empirical evidence is not likely to be forthcoming quickly. Thus it becomes a matter of some importance to establish criteria for cautious generalization of findings between invertebrate and vertebrate groups.

A number of approaches to this problem of generalization are possible. One important approach involves evaluating known similarities and differences in the structure, function, and biochemistry of insect and vertebrate nervous systems. However, perhaps an equally important approach involves examining and comparing the behavior of the invertebrate preparation with the behavior of vertebrates in various learning situations. Thus a brief analysis of the theoretical status of learning in headless insects will be presented here.

The learning mediated by the insect ventral nerve cord and its associated ganglia has often been called avoidance learning (e.g., Eisenstein, 1970a, b, 1971). Although many theories of avoidance learning have been proposed, the most widely accepted theory developed to explain avoidance learning in vertebrates (Mowrer, 1947; Solomon and Wynne, 1954) holds that it involves two stages: (1) classical conditioning in which the CS is some stimulus which reliably precedes onset of the noxious UCS which the organism ultimately learns to avoid and (2) instrumental escape learning for which the reinforcement is initially escape from the noxious UCS but later escape from the conditioned "fear" response elicited by the CS. According to this view, then, avoidance consists of escaping from the CS before the UCS comes on.

Although they do not explicitly indicate adherence to such a view, statements by Horridge (1962a,b) and Eisenstein (e.g., 1971) to the effect that "P" animals form an association between shock and a particular leg position constitute tacit acceptance of at least the classical conditioning portion of such a two-process avoidance learning theory. Moreover, if the leg flexion learning of headless insects is to be viewed as analogous to the avoidance learning of vertebrates, performance of the avoidance response must be associated with *some* stimulus, and, in the absence of any other obvious sensory channels through which such a stimulus could be transmitted, proprioceptive feedback stimuli appear to be the obvious choice. Nevertheless, Hoyle (1965) demonstrated that an electrophysiological analogue of leg flexion learning could occur in headless locusts in which proprioceptive receptors had been dissected away from the relevant muscles. Thus serious doubt is cast on the necessity of the stimuli, which must be presumed to become associated with shock if the learned leg flexion behavior of headless insects is to be viewed as analogous to vertebrate avoidance learning.

It is of course possible, as Eisenstein (1971) implies, that there is a species difference between locusts and cockroaches. Perhaps in cockroaches neither behavioral leg flexion learning nor its electrophysiological analogue would occur without proprioceptive feedback. However, such a species difference in insects belonging to closely related orders seems unlikely. Moreover, if such a basic difference were found between species as closely related as cockroaches and locusts, this finding would be reason for pessimism about the possibility of generalizing from cockroaches to vertebrates.

However, the presence or absence of this species difference is a matter which should be determined empirically by attempting to replicate Hoyle's (1965) experiment with the cockroach. The experiment by Murphy (1967) cited by Eisenstein (1971) is inconclusive in that only the hair plates were removed.

Moreover, in my opinion, an experiment by Disterhoft *et al.* (1971), the results of which are interpreted by these authors in associative terms, is in fact equally open to interpretation in nonassociative terms. These investigators first gave simulated "R" training (i.e., non–response contingent leg shock) to headless specimens of *Periplaneta americana,* the shocked leg being either bound in a flexed position, bound in an extended position, or free to move about. All animals then received "P" training, with shock administered whenever the leg fell below a predetermined level. The results during the second phase of the study were that the animals whose legs had been free to move took the most shock, while animals whose legs had been bound in the extended position took the fewest, the performance of the animals whose legs had been bound in the flexed position being intermediate between that of the other two groups. The author agrees with Disterhoft *et al.* (1971) that these findings offer strong support for Eisenstein's (1971; Eisenstein and Cohen, 1965) contention that an animal whose leg is free to move about learns many competing *responses* during "R" training. However, superiority of the extended condition over the flexed condition offers little support for the view that associations are formed between propriocepted leg positions and shock. The difference between the extended and flexed groups can be explained by the alternative hypothesis that, after unbinding, a previously flexed leg would show a compensatory tendency to adopt a somewhat extended position, while a previously extended leg would show the opposite tendency. Indeed, if there were any great tendency to associate leg position with shock, one would expect that animals shocked in the flexed position would have formed such a strong association between that position and the shock that flexion learning would be more difficult for them than for unbound control animals.

Thus at this time it appears likely that leg flexion learning in cockroaches is not equivalent to vertebrate avoidance learning, because of the absence of any association between the shock stimulus and another stimulus. For this reason, it would appear to be useful to examine an alternative to the avoidance learning interpretation. Such an alternative is offered by Razran's (1971) conclusion that leg flexion learning in headless insects is aversive inhibitory conditioning, a form of learning less complex than either classical or instrumental conditioning and thus probably evolutionarily more primitive.

Razran (1971, p. 17) defines aversive inhibitory conditioning as a decrement or disappearance of a reaction through an associated antagonistic reaction." My view[5] is that aversive inhibitory learning in headless insects involves inhibition of two reactions. The first of these is the resting animal's tendency to maintain its leg in a relatively relaxed and moderately extended position. The second is a part of the unconditioned reaction to shock. Shock initially elicits a complex unconditioned response involving both extension and flexion components, although flexion apparently predominates (Eisenstein, 1971). In the standard "P" learning situation, both the resting posture and the extension components of the unconditioned reaction to shock are punished and inhibited, thereby strengthening leg flexion. The learning assumed to be involved here is not instrumental avoidance learning, since no "new" response is learned, the learned response being a part of the unconditioned response. And, to reiterate, it is not classical conditioning, since no other stimulus is presumed to become associated with the shock.

Pritchatt (1968) showed that *intact* cockroaches could also be trained to extend their legs, and extension learning may also be possible in headless insects, although this has not yet been shown. Such a finding would mean that the flexion components of the unconditioned response are also capable of being inhibited by punishment. However, another possibility is that Pritchatt's intact animals were showing genuine avoidance learning and that extension learning will prove to be impossible for headless animals.

In his review of aversive inhibitory conditioning, Razran (1971) cites examples of the phenomenon in animals ranging from hydra to man. It is thus phylogenetically an extremely widespread means of behavior modification, more widespread than classical or instrumental conditioning and certainly more widespread than avoidance learning, which involves an interaction of the latter two. For that reason, leg flexion learning in headless insects viewed as aversive inhibitory conditioning may perhaps be regarded as a phenomenon whose molecular correlates are more likely to be similar in insects and vertebrates than are the molecular correlates of avoidance learning. Avoidance learning may have evolved in parallel or convergent fashion in the higher invertebrates and in the vertebrates. However, aversive inhibitory conditioning is a capacity which is more likely to have been possessed by the common ancestor of these two very different animal groups.

X. OLFACTORY CONDITIONING

Another set of phenomena which resembles learning in insects was discovered by Thorpe and Jones (1937). These investigators tested female

[5]Inasmuch as I disagree with one example of aversive inhibitory conditioning cited by Razran (1971), I am unsure whether Razran would agree with the details of the present analysis of leg flexion learning in the cockroach.

parasitic wasps *(Nemeritis canescens)* in an olfactometer. Females reared on their species' normal host, the Mediterranean flour moth *(Ephestia kuhniella)*, were strongly attracted to the odor of *Ephestia* but indifferent to the odor of the wax moth *(Meliphora grisella)*. However, *Nemeritis* females which had been reared artificially on *Meliphora* larvae showed a tendency to approach the odor of *Meliphora*, and some of them even chose the odor of *Meliphora* over that of *Ephestia*. Thus rearing *Nemeritis* females on an abnormal host modified an olfactory response which plays an important role in the selection of a site for oviposition in this species.

Thorpe (1939*a*) showed further that this "conditionability" of olfactory approach responses is not limited to the preimaginal stages of life. *Nemeritis* adults also show a tendency to approach both the odor of *Meliphora* and that of cedar wood oil after exposure to these odors. Similar phenomena have also been demonstrated with fruit flies (*Drosophila melanogaster* and *D. guttifera*), which come to approach the odor of peppermint oil (a normally repulsive odor) after being reared in a peppermint-scented culture medium (Thorpe, 1939*b*; Cushing, 1941).

Thorpe (1943) interpreted the acquired attractiveness of peppermint in *Drosophila* as an example of habituation. It was postulated that oil of peppermint contains two substances, one of which is attractive and the other repulsive to *Drosophila*. Exposure was thought to produce habituation to the repulsive substance and thereby allow the attraction to manifest itself.

Somewhat different mechanisms have been invoked to explain the acquired attractiveness of the odor of *Meliphora* to *Nemeritis*. It will be recalled that the odor of *Meliphora* is normally a neutral one for *Nemeritis* so that an explanation in terms of habituation of some negative component is unsatisfactory. Thorpe (1939*a*) accounted for the acquired attractiveness by assuming that it was the result of the generally good environment in which the subjects lived while being exposed to it. This hypothesis has received some recent experimental support from a study (Hershberger and Smith, 1967) which showed that peppermint loses its attractiveness for "conditioned" *Drosophila* subjects if they are confined in its presence without food. Thorpe (1943) has also explained the "conditioned" attractiveness of *Meliphora* to *Nemeritis* as being an example of latent learning. However, this latter explanation seems to amount to little more than substitution of one name for another.

One common definition of learning states that learning is any relatively permanent change in behavior which results from practice (e.g., Kimble, 1961). If this definition is accepted, then the meaning of "practice" becomes a matter of major importance in considering whether "olfactory conditioning" is a form of learning. In the other examples of learning which have been considered in this chapter, practice could be taken to mean the performance of at least some approximation of the responses whose frequency of occurrence learning affects. Yet, in olfactory conditioning, the animal simply lives in an environment permeated with an odor and then shows an approach response to that odor without having had any obvious practice in making similar approach responses. Thus, if practice is viewed as being crucial to the definition of learning, olfactory conditioning appears to be a very dubious candidate for inclusion in the category of learned behavior changes. However, preferences concerning definitions of learning vary, and since olfactory conditioning clearly involves an effect of experience on behavior, it is related to learning in at least a general way.

XI. PROCESSES OF LEARNING AND MEMORY

A. Learning

Schneirla (1929, 1941, 1943, 1962) has shown that the learning of complex mazes by ants can be characterized as consisting of at least two and possibly three stages. The first stage is nonspecific in the sense that no tendencies toward the elimination of specific blind alleys are in evidence. Instead, the first stage involves the rapid diminution of tendencies to run in circles, to crawl on the walls and ceiling of the maze, and to make repeated entries into blind alleys on the same trip through the maze. This first stage is said to last for about eight trials (Schneirla, 1941). In the second stage, the ant learns to follow the true path and to avoid entrances into individual blind alleys. How long this stage lasts depends on the number of choice points in the maze and centrifugal swing (Schneirla, 1943). Finally, a third stage, the existence of which is less clearly documented, may involve integration of the maze habit into a more-unitary, smoothly running, and possibly kinesthetically controlled unit (Schneirla, 1929, 1962). Vowles has confirmed the existence of the first two stages of ant learning in a maze situation much simpler than that employed by Schneirla (1929, 1941, 1943). However, whether these stages of maze learning are characteristic of all insects or just of ants is unclear.

Another feature of insect learning is an apparently high degree of situation specificity; what is learned in one context is not easily applied in another. After determining that the same complex maze pattern is learned more quickly when ants traverse the maze on their way back to the nest than when the ants traverse the maze on their way to the food source (Schneirla, 1933), Schneirla (1934) examined transfer of maze learning from one leg of a foraging trip to the other. Ants ran a maze on their way to the food source and returned over a direct route until the maze had been thoroughly learned. They then learned the identical maze pattern on return trips to the nest, outward-bound trips being over the direct route. A second group of ants experienced the two phases of the experiment in reverse order. And a third group of ants ran through the two mazes which presented the same sequence of turns on both outward-bound and return trips. The findings were (1) that learning the maze on one leg of a foraging trip in no way facilitated subsequent learning of the same maze pattern on the opposite leg of such trips and (2) that almost no learning at all occurred when ants were confronted with the same sequence of turns on both legs of foraging trips. Weiss and Schneirla (1967) confirmed the fact that the specific responses learned on one leg of a foraging trip do not transfer to trips made in the opposite direction.

A very different kind of evidence of the high degree of situation specificity in insect learning was obtained by Longo (1964). This investigator trained cockroaches to turn either right or left in a T-maze and then repeatedly reversed the discrimination. No evidence was found to indicate that experience in learning reversals led to any improvement in the animals' ability to learn reversals. Except for a slight general adaptation to the maze situation which was evidenced also by roaches exposed to their first reversal after an equivalent amount of training to only one side, the animals still behaved as though each new reversal had to be learned as if it were the first reversal experienced. In contrast, in studies involving the use of single-choice-point mazes with ants, Schneirla (1960, 1962) found evidence for improvement with successive reversals of position discriminations. However, numerous procedural differences between the Longo (1964) and Schneirla (1960, 1962) studies make the significance of this seeming difference between ants and cockroaches difficult to assess. If it could be replicated under more comparable conditions, this difference between very advanced (ants) and rather primitive (cockroaches) insects would be comparable to a similar difference between mammals and fish observed by Bitterman and his associates (e.g., Bitterman, 1960, 1965; cf. Brookshire, 1970, for a recent review).

Another aspect of Longo's (1964) study dealt with an instrumental conditioning procedure known as probability learning. In such a situation, the animal is confronted with alternatives (the two sides of a T-maze in this case), both of which are sometimes reinforced, but with different probabilities. Problems were investigated in which the reinforcement probabilities were 100:0, 70:30, 60:40, and 50:50. It was found that the cockroaches matched their probability of choice of an alternative to the probability of its being reinforced; thus a 70% alternative was chosen 70% of the time. In this situation also, the cockroach's behavior differs from that of non-human mammals and resembles that of fish (Bitterman, 1965; Brookshire, 1970).

B. Memory

The widely held interference theory of memory and forgetting maintains that forgetting is caused by activity (both overt and neural) which interferes with retention and not by any passive decay process. An implication of this view is that anything which reduces overt or neural activity should preserve memory, while anything which increases such activity should be destructive of memory. Two means of lowering or raising activity levels in insects are mechanical stimulation and restraint and exposure to high and low temperatures. Apparently, both methods produce the predicted effects. Minami and Dallenbach (1946) trained cockroaches *(Periplaneta americana)* on the Szymanski (1912)–Turner (1912) shock avoidance task and showed that enforced treadmill running administered immediately after learning interfered with retention, while enforced inactivity produced by wedging subjects between folds of tissue paper facilitated retention. Hoagland (1931) found that exposure to high temperatures of 29.4°C interfered with both acquisition and retention of shuttle box escape learning in the ant *(Componotus herculeanus pennsylvanicus)*. And Alloway (1969, 1970; Alloway and Routtenberg, 1967) has shown that exposure to cold (1.7°C) between daily training sessions and during retention intervals of 1–10 days facilitates both acquisition and retention of maze learning in the grain beetle *(Tenebrio molitor)*. Two studies which seem to show an inhibitory effect of cold on learning (van der Heyde, 1920; Hunter, 1932) and retention (Hunter, 1932) can probably be discounted. In one case (Hunter, 1932), the low temperatures employed were probably injurious to the cockroach subjects. In the other case (van der Heyde, 1920), the animals were exposed to low temperatures during training sessions and were thus likely to have been impaired in their ability to make the correct response (jumping).

Other evidence suggests that insect memory may be a two-stage process involving a short-term consolidation phase before the memory becomes permanently fixated. The first evidence suggesting this possibility was the finding of Minami and Dallenbach (1946) that treadmill running interferes with retention in the cockroach only when it is administered immediately after learning. Treadmill running administered after an hour of normal activity following learning has no effect on retention.[6] This fact suggests that memory is particularly vulnerable to interference for a short time after learning. Further support for this notion is provided by the fact that the administration of carbon dioxide anesthesia[7] immediately after each learning

[6] A tendency for treadmill running administered immediately before training sessions to interfere with both original learning and relearning tests of retention is attributable to fatigue.

[7] Exposure to carbon dioxide produces a convulsion before inducing anesthesia. Thus it should be viewed as a means of raising (not lowering) activity levels.

trial prevents shuttle box shock avoidance learning in cockroaches (Freckleton and Wahlsten, 1968). Plath's (1924) finding that ether anesthesia does not cause bees to forget the location of their hive is consistent with the view that memory for well-learned and presumably permanently stored information may not be similarly vulnerable.

Another line of evidence suggesting a two-phase memory consolidation process in insects is studies of retention revealing a period of poor retention preceded and followed by periods of good retention. For example, Eisenstein (1970b), in a study of retention of leg flexion learning in headless cockroaches, found deteriorating retention in tests administered 10 min, 30 min, and 1 hr after training followed by improving performance on tests administered 2 and 12 hr after training. Similarly, Alloway (1969, 1970; Alloway and Routtenberg, 1967), in a series of studies of the effects of cold on retention of maze learning in grain beetles *(Tenebrio molitor)*, has found that, although beetles exposed to cold during retention intervals of 1, 3, 4, 5, and 10 days show better retention than animals left at room temperature for similar periods, animals tested after a cold 2-day retention interval show almost no retention. Walrath (1970) has also observed better retention after cold 1-, 3-, and 4-day retention intervals in *Tenebrio* beetles. Moreover, the poor performance of the cold 2-day animals cannot be attributed to nonspecific motivational or performance factors (Alloway and Routtenberg, 1967) or to a reversal of turning bias (Routtenberg et al., 1968). Instead, a genuine loss of memory produced by warming at a critical period 48 hr after training seems to be involved (Alloway, 1970). Although other interpretations are possible, these findings suggest (1) that cold slows the memory consolidation process and (2) that the poor performance after the cold 2-day retention interval may result from disturbance of consolidation at some crucial point such as the transfer of the "engram" from temporary into permanent storage.

XII. OVERVIEW AND SUGGESTIONS FOR FUTURE RESEARCH

In considering the experimental evidence reviewed in this chapter as a whole, two major points should be made. First, the findings of Horridge (1962a,b), Eisenstein and Cohen (1965), Hoyle (1965), and Kerkut et al. (1970) together with the rest of the work on leg flexion learning in headless insects represent an immense advance toward the goal of isolating the site where a simple behaviorally measurable form of learning occurs and then determining the nature of the neurochemical processes which produce it. This work would appear to be comparable in importance to the work with cellular analogues of learning in *Aplysia* and the studies of memory transfer in *Planaria,* which are discussed in other chapters of this book.

However, despite the importance and promise of the work with headless insect preparations, one cannot escape the conclusion that our behavioral knowledge about insect learning is meager indeed. Moreover, this dearth of behavioral information is serious inasmuch as it impedes attempts to determine what relevance the advances which are being made at more molecular levels may have to vertebrate neurophysiology and the understanding of vertebrate learning.

The lack of behavioral information about learning in insects is obvious in many ways. For example, a number of the basic phenomena of classical and instrumental conditioning have not yet been demonstrated with insects, and the effects of many of the important independent variables known to influence conditioning in vertebrates have never been examined in insects. Attempts to demonstrate such basic phenomena as extinction, spontaneous recovery, and stimulus generalization are urgently needed in both conditioning paradigms, and so too are studies of the effects of such important independent variables as reinforcement magnitude in instrumental conditioning and length of the CS–UCS interval in classical conditioning.

Moreover, there is a serious lack of comparative research, and the information which we do possess is fragmentary. Almost all the behavioral learning studies have been performed with cockroaches (Dictyoptera), grain beetles (Coleoptera), fruit flies and blowflies (Diptera), and ants, wasps, and bees (Hymenoptera), i.e., with representatives of four of 28 orders. No studies of learning have been published with insects from the phylogenetically interesting apterygotan (primitively wingless) orders, and there have been only a few studies of learning in those groups of arthropods and annelids which are thought to resemble the animals from which insects evolved.

In addition, few comparisons are possible even among the more commonly studied insect groups because comparable studies of the same phenomena are rarely available. Perhaps the most striking example of this problem is the fact that instrumental learning has yet to be demonstrated in the blowfly, the only insect for which we have a convincing demonstration of classical conditioning.

These considerations have led me to believe that a considerable amount of effort is going to have to be expended in broadening our knowledge of the molar behavioral laws of learning in a phylogenetically representative sample of insects and other invertebrates before it will be possible to exploit fully the current advances in the neurophysiology and biochemistry of learning in simplified preparations. Thus systematic studies of the effects of vertebrate-type independent variables are needed to facilitate comparisons with the more commonly studied vertebrate groups, and systematic comparisons among animals representative of the broad evolutionary differences within the insect class and between that class and its close arthropod and annelid relatives are needed to trace the evolution of learning capacities in insects.

REFERENCES

Alloway, T. M., 1969, Effects of low temperature upon acquisition and retention in the grain beetle, Tenebrio molitor, J. Comp. Physiol. Psychol., 69, 1–8.

Alloway, T. M., 1970, Methodological factors affecting the apparent effects of exposure to cold upon retention in the grain beetle, Tenebrio molitor, J. Comp. Physiol. Psychol., 72, 311–317.

Alloway, T. M., 1972, Retention of learning through metamorphosis in the grain beetle, Tenebrio molitor, Am. Zoologist, 12, 471–477.

Alloway, T. M., and Routtenberg, A., 1967, "Reminiscence" in the cold flour beetle (Tenebrio molitor), Science, 158, 1066–1067.

Aranda, L. C., and Luco, J. V., 1969, Further studies of an electric correlate to learning: Experiments in an isolated insect ganglion, Physiol. Behav., 4, 133–137.

Baxter, C., 1957, Habituation of the roach to puffs of air, Anat. Rec., 128, 521 (abst.).

Bitterman, M. E., 1960, Towards a comparative psychology of learning, Am. Psychologist, 15, 704–712.

Bitterman, M. E., 1965, The evolution of intelligence, Sci. Am., 212, 92–100.

Borsellino, A., Pieratoni, R., and Schietti-Cavazza, B., 1967, Persistenza nello stadio adulto di Tenebrio molitor (coleottero tenebrionide) de comportamenti indotti allo stato larvale, Atti Congr. Ann., Grouppo Naz. Cibernet. C.N.R., Pisa, pp. 29–36.

Borsellino, A., Pieratoni, R., and Schietti-Cavazza, B., 1970, Survival in adult mealworm beetles (Tenebrio molitor) of learning acquired at the larval stage, Nature (Lond.), 225, 963–964.

Brookshire, K. H., 1970, Comparative psychology of learning, in "Learning: Interactions" (M. H. Marx, ed.), pp. 290–364, Collier-Macmillan, Toronto.

Brown, B. M., and Noble, E. P., 1967, Cycloheximide and learning in the isolated cockroach ganglion, Brain Res., 6, 363–366.

Brown, B. M., and Noble, E. P., 1968, Cycloheximide, amino acid incorporation, and learning in the isolated cockroach ganglion, Biochem. Pharmacol., 17, 2371–2374.

Bullock, T. H., and Horridge, G. A., 1965, "Structure and Function in the Nervous Systems of Invertebrates," W. H. Freeman, London and San Francisco.

Chapman, R. F., 1969, "The Insects: Structure and Function," English Universities Press, London.

Chauvin, R., 1947, Études sur le comportement de Blattella germanica dans divers types de labyrinthes, Bull. Biol. France Belg., 61, 92–128.

Cherkashin, A. N., Sheyman, E. M., and Stafekhina, V. S., 1968: O sokhanenii vremennikh svyazei oo nasekhomikh v protsesse metamorfoza (Conservation of conditioned reflexes during metamorphosis in insects), Zh. Evol. Biochim. Fiziol., 117–120.

Cushing, J. E., Jr., 1941, An experiment on olfactory conditioning in Drosophila guttifera, Proc. Nat. Acad. Sci., 50, 163–178.

Dethier, V. G., 1969, Feeding behavior of the blowfly, in "Advances in the Study of Behavior" (D. S. Lehrman, R. A. Hinde, and E. Shaw, eds.), Vol. 2, Academic Press, New York.

Dethier, V. G., Solomon, R. L., and Turner, L. H., 1965, Sensory input and central excitation and inhibition in the blowfly, J. Comp. Physiol. Psychol., 60, 303–313.

Disterhoft, J. F., Nurnberger, J., and Corning, W. C., 1968, "P–R" differences in intact cockroaches as a function of testing interval, Psychon. Sci., 12, 205–206.

Disterhoft, J. F., Haggerty, R., and Corning, W. C., 1971, An analysis of leg position learning in the cockroach yoked control, Physiol. Behav., 7, 1026–1029.

Ebeling, W., Wagner, R. E., and Rierson, D. A., 1966, Influence of repellency on the efficacy of blatticides: I. Learned modification of behavior of the German cockroach, *J. Econ. Entomol., 59,* 1374–1388.

Edwards, J. S., 1969, Postembryonic development and regeneration of the insect nervous system, *in* "Advances in Insect Physiology" (J. W. L. Beament, J. E. Treherne, and V. B. Wigglesworth, eds.), Vol. 6, pp. 97–137, Academic Press, London.

Edwards, J. S., 1970, Neural control of development in arthropods, *in* "Invertebrate Nervous Systems: Their Significance for Mammalian Neurophysiology" (C. A. G. Wiersma, ed.), pp. 95–109, University of Chicago Press, Chicago.

Eisenstein, E. M., 1965, Effects of strychnine sulfate and sodium pentobarbitol on shock-avoidance learning in an isolated insect ganglion, *Proc. Seventy-third Ann. Meeting Am. Psychol. Ass.,* pp. 127–128.

Eisenstein, E. M., 1968, Assessing the influence of pharmacological agents on shock avoidance in simpler systems, *Brain Res., 11,* 471–480.

Eisenstein, E. M., 1970a, A comparison of activity and position response measures of avoidance learning in the cockroach, *Periplaneta americana, Brain Res., 21,* 143–147.

Eisenstein, E. M., 1970b, The retention of shock avoidance learning in the cockroach, *Periplaneta americana, Brain Res., 21,* 148–150.

Eisenstein, E. M., 1971, Learning in isolated insect ganglia: A survey of procedures and results including some theoretical speculations and suggested experiments, *in* "Experiments in Physiology and Biochemistry" (G. Kerkut, ed.), Vol. 2, Academic Press, New York.

Eisenstein, E. M., and Cohen, M. J., 1965, Learning in an isolated insect ganglion, *Anim. Behav., 13,* 104–108.

Eldering, F. J., 1919, Acquisition d'habitudes par les insectes, *Arch. Néerl. Physiol., 3,* 469–490.

Evans, S., 1932, An experiment in maze learning in ants, *J. Comp. Psychol., 14,* 183–190.

Fox, R. M., and Fox, J. W., 1964, "Introduction to Comparative Entomology," Reinhold, New York.

Freckleton, W. C., and Wahlstein, D., 1968, Carbon dioxide induced amnesia in the cockroach, *Psychon. Sci., 12,* 179–180.

Frings, H., 1941, The loci of olfactory end-organs in the blow-fly, *Cynomyia cadaverina* Desvoidy, *J. Exptl. Zool., 88,* 65–93.

Gates, M. F., and Allee, W. C., 1933, Conditioned behavior of isolated and grouped cockroaches on a simple maze, *J. Comp. Psychol., 15,* 331–358.

Glassman, E., Henderson, A., Cordle, M., Moon, H. M., and Wilson, J. E., 1970, Effect of cycloheximide and actinomycin D on the behaviour of the headless cockroach, *Nature (Lond.), 225,* 967–968.

Goustard, M., 1948, Inhibition de la photonégativité par le dressage, chez *Blattella, Compt. Rend. Acad. Sci. (Paris), 227,* 785–786.

Hershberger, W. A., and Smith, M. P., 1967, Conditioning in *Drosophila melanogaster, Anim. Behav., 15,* 259–262.

Hoagland, H., 1931, A study of the physiology of learning in ants, *J. Gen. Psychol., 5,* 21–41.

Hollis, J. H., 1963, Habituatory response decrement in pupae of *Tenebrio molitor, Anim. Behav., 11,* 161–163.

Horn, G., and Rowell, C. H. F., 1968, Medium and long-term changes in the behaviour of visual neurones in the tritocerebrum of locusts, *J. Exptl. Biol., 49,* 143–169.

Horridge, G. A., 1962a, Learning of leg position by headless insects, *Nature (Lond.), 193,* 697–698.

Horridge, G. A., 1962b, Learning of leg position by the ventral nerve cord in headless insects, *Pro. Roy. Soc. (Lond.), B157,* 33–52.

Horridge, G. A., Sholes, J. H., Shaw, S., and Tunstall, J., 1965, Extracellular recordings from single neurons in the optic lobe and brain of the locust, *in* "The Physiology of the Insect Central Nervous System" (J. E. Treherne and J. W. L. Beament, eds.), pp. 165–202, Academic Press, New York.

Hoyle, G., 1965, Neurophysiological studies on "learning" in headless insects, *in* "The Physiology of the Insect Central Nervous System" (J. E. Treherne and J. W. L. Beament, eds.), pp. 203–232, Academic Press, New York.

Hoyle, G., 1970, Cellular mechanisms underlying behavior—Neuroethology, *in* "Advances in Insect Physiology" (J. W. L. Beament, J. E. Treherne, and V. B. Wigglesworth, eds.), Vol. 7, pp. 394–444, Academic Press, London.

Hullo, A., 1948, Rôle des tendances motrices et des données sensorielles dans l'apprentissage du labyrinthe par les blattes (*Blattella germanica*), *Behaviour, 1*, 297–310.

Hunter, W. S., 1932, The effect of inactivity produced by cold upon learning and retention in the cockroach, *Blattella germanica, J. Genet. Psychol., 41*, 253–266.

Imms, A. D., 1957, "A General Textbook of Entomology," 9th ed., Methuen, London.

Kandel, E., Castellucci, V., Pinsker, H., and Kupfermann, I., 1970, The role of synaptic plasticity in the short-term modification of behaviour, *in* "Short-Term Changes in Neural Activity and Behaviour" (G. Horn and R. A. Hinde, eds.), pp. 281–322, Cambridge University Press, Cambridge.

Kerkut, G. A., Oliver, G., Rick, J. T., and Walker, R. J., 1970, Biochemical changes during learning in an insect ganglion, *Nature (Lond.), 227*, 722–723.

Kimble, G. A., 1961, "Hilgard and Marquis' Conditioning and Learning," 2nd ed., Appleton-Century-Crofts, New Yrok.

Lanham, U., 1964, "The Insects," Columbia University Press, New York.

Le Bigot, L., 1952, Réminiscence au cour d'un apprentissage moteur en fonction de la répétition de l'exercise chez *Blattella germanica, Rev. Sci.*, pp. 254–258.

Le Bigot, L., 1954, Influence des sanctions sur l'acquisition et la retention chez *Blattella germanica, J. Psychol. Norm. Pathol., 51*, 327–335.

Lecompte, J., 1948, Recherches sur les faiteurs de la motivation chez *Blattella germanica, Bull. Soc. Zool. France, 83*, 215–220.

Longo, N., 1964, Probability learning and habit reversal in the cockroach, *Am. J. Psychol., 77*, 29–41.

Longo, N., 1970, A runway for the cockroach, *Behav. Res. Methods Instrumentation, 2*, 118–119.

Manton, S. M., 1964, Mandibular mechanisms and the evolution of arthropods, *Phil. Trans. Roy. Soc. (Lond.), 247*, 5–183.

Minami, H., and Dallenbach, K. M., 1946, The effect of activity upon learning and retention in the cockroach, *Periplaneta americana, Am. J. Psychol., 59*, 1–58.

Mowrer, O. H., 1947, On the dual nature of learning—A re-interpretation of "conditioning" and "problem-solving," *Harvard Educ. Rev., 17*, 102–148.

Murphy, R. K., 1967, Proprioception and learning of leg position in cockroaches, Unpublished masters thesis, cited by Eisenstein (1971).

Murphy, R. M., 1967, Instrumental conditioning of the fruit fly, *Drosophila melanogaster, Anim. Behav., 15*, 153–161.

Murphy, R. M., 1969, Spatial discrimination performance of *Drosophila melanogaster*: Some controlled and uncontrolled correlates, *Anim. Behav., 17*, 43–46.

Nelson, M. C., 1971, Classical conditioning in the blowfly (*Phormia regina*): Associative and excitatory factors, *J. Comp. Physiol. Psychol., 77*, 353–368.

Niklaus, R., 1965, Die Erregung einzelner Fadenhaare von *Periplaneta americana* in Abhängikeit von der Grösse und Richtung der Anstenkung, *Z. Vergl. Physiol., 50*, 331–362.

Plath, O. E., 1924, Do anesthetized bees lose their memory? *Am. Naturalist, 58,* 162–166.

Pritchatt, D., 1968, Avoidance of electric shock by the cockroach, *Periplaneta americana, Anim. Behav., 16,* 178–185.

Pumphrey, R. J., and Rawdon-Smith, A. F., 1936, Synchronized action potentials in the cercal nerve cord of the cockroach (*Periplaneta americana*) in response to auditory stimuli, *J. Physiol., 87,* 4p–5p.

Pumphrey, R. J., and Rawdon-Smith, A. F., 1937, Synaptic transmission of nervous impulses through the last abdominal ganglion of the cockroach, *Proc. Roy. Soc. (Lond.), B122,* 106–118.

Rabaud, E., 1926, Acquisition des habitudes et repères sensoriels chez les guepes, *Bull. Sci. France Belg., 60,* 313–333.

Razran, G. A., 1971, "Mind in Evolution: An East–West Synthesis," Houghton-Mifflin, Boston.

Ross, H. H., 1965, "A Textbook of Entomology," 3rd ed., Wiley, New York.

Routtenberg, A., Alloway, T. M., and Hill, W, F., 1968, Reply to Deutsch, *Science, 160,* 1024.

Rowell, C. H. F., 1970, Incremental and decremental processes in the insect central nervous system, *in* "Short-Term Changes in Neural Activity and Behaviour" (G. Horn and R. A. Hinde, eds.), pp. 237–280, Cambridge University Press, Cambridge.

Rowell, C. H. F., and Horn, G., 1968, Dishabituation and arousal in the response of single nerve cells in an insect brain, *J. Exptl. Biol., 49,* 171–183.

Schneirla, T. C., 1929, Learning and orientation in ants, *Comp. Psychol. Monogr., 6,* No. 4.

Schneirla, T. C., 1933, Motivation and efficiency in ant learning, *J. Comp. Psychol., 15,* 243–266.

Schneirla, T. C., 1934, The process and mechanism of ant learning: The combination-problem and the successive-presentation problem, *J. Comp. Psychol., 17,* 303–328.

Schneirla, T. C., 1941, Studies on the nature of ant learning: I. The characteristics of a distinctive initial period of generalized learning. *J. Comp. Psychol., 32,* 41–82.

Schneirla, T. C., 1943, The nature of ant learning: II. The intermediate stage of segmental maze adjustment, *J. Comp. Psychol., 34,* 149–176.

Schneirla, T. C., 1953, Modifiability in insect behavior, *in* "Insect Physiology" (K. D. Roeder, ed.), pp. 723–747, Wiley, New York.

Schneirla, T. C., 1960, L'apprentissage et la question du Conflit chez la fourni—comparaison avec le rat, *J. Psychol., 57,* 11–44.

Schneirla, T. C., 1962, Psychological comparison of insect and mammal, *Psychol. Beitrage, 6,* 509–520.

Shepard, J., 1911, Some results in comparative psychology, *Psychol. Bull., 8,* 41–42.

Smith, D. S., and Treherne, J. E., 1963, Functional aspects of the organization of the insect nervous system, *in* "Advances in Insect Physiology" (J. W. L. Beament, J. E. Treherne, and V. B. Wigglesworth, eds.), Vol. 1, pp. 401–484, Academic Press, London.

Snodgrass, R. E., 1938, Evolution of Annelida, Onychophora, and Arthropoda, *Smithsonian Misc. Coll., 97,* 1–159.

Solomon, R. L., and Wynne, L. C., 1954, Traumatic avoidance learning: The principles of anxiety conservation and partial irreversibility, *Psychol. Rev., 61,* 353–385.

Szymanski, J. S., 1912, Modification of the innate behavior of cockroaches, *J. Anim. Behav., 2,* 81–90.

Thorpe, W. H., 1939a, Further studies on olfactory conditioning in a parasitic insect. The nature of the conditioning process, *Proc. Roy. Soc. (Lond.), B126,* 379–397.

Thorpe, W. H., 1939b, Further studies of pre-imaginal olfactory conditioning in insects, *Proc. Roy. Soc. (Lond.), B127,* 424–433.

Thorpe, W. H., 1943, Types of learning in insects and other arthropods, *Brit. J. Psychol.*, *33*, 220–234; *34*, 20–31, 66–76.

Thorpe, W. H., and Jones, F. G. W., 1937, Olfactory conditioning and its relation to the problem of host selection, *Proc. Roy. Soc. (Lond.)*, *B124*, 56–81.

Tiegs, O. W., and Manton, S. M., 1958, The evolution of Arthropoda, *Biol. Rev.*, *33*, 255–332.

Turner, C. H., 1912, An experimental investigation of an apparent reversal of responses to light of the roach (*Periplaneta orientalis* L.), *Biol. Bull.*, *23*, 371–386.

Turner, C. H., 1913, Behavior of the common roach (*Periplaneta orientalis* L.) on an open maze, *Biol. Bull.*, *25*, 348–365.

van der Heyde, H. C., 1920, Quelques observations sur la psychologie des fournis, *Arch. Néerl. Physiol.*, *4*, 259–282.

Verlaine, L., 1924, L'instinct et l'intelligence chez les hymenoptères: III. La reconnaissance du nid et l'éducabilité de l'odorate chez la *Vespa germanica* Fab., *Ann. Soc. Roy. Entomol. Belg.*, *15*, 67–117.

Verlaine, L., 1925, L'instinct et l'intelligence chez les hymenoptères: V. La traversée d'un labyrinthe par des guèpes et des bourdons (*Vespa germanica* Linn., *V. crabro* Linn., *Bombus terrestris* Linn., et *B. sylvarum* Linn.), *Ann. Soc. Roy. Zool. Belg.*, *56*, 33–98.

Verlaine, L., 1927, L'instinct et l'intelligence chez les hymenoptères: VII. L'abstraction, *Bull. Ann. Soc. Entomol. Belg.*, *58*, 59–88.

von Borell du Vernay, W., 1942, Assozionsbildung und Sensibilisierung bei *Tenebrio molitor* L., *Z. Vergl. Physiol.*, *30*, 84–116.

Vowles, D. M., 1964, Olfactory learning and brain lesions in the wood ant (*Formica rufa*), *J. Comp. Physiol. Psychol.*, *58*, 105–111.

Vowles, D. M., 1965, Maze learning and visual discrimination in the wood ant (*Formica rufa*), *Brit. J. Psychol.*, *56*, 15–31.

Walrath, L. C., 1970, Retention and interference in the beetle, *Tenebrio molitor*, *Psychon. Sci.*, *18*, 267–268.

Warden, C. J., Jenkins, T. N., and Warner, L., 1941, "Comparative Psychology: A Comprehensive Treatment," Vol. II: "Plants and Invertebrates," Ronald Press, New York.

Weiss, B. A., and Schneirla, T. C., 1967, Inter-situational transfer in the ant *Formica schaufussi* as tested in a two-phase single choice point maze, *Behaviour*, *28*, 269–279.

Wheeler, W. M., 1928, "The Social Insects," Harcourt, Brace, New York.

Zajonc, R. B., Heingartner, A., and Herman, E. M., 1969, Social enhancement and impairment of performance in the cockroach, *J. Personal. Soc. Psychol.*, *13*, 83–92.

Chapter 9

HONEY BEES

Patrick H. Wells

Department of Biology
Occidental College
Los Angeles, California, USA

I. INTRODUCTION

> Foraging behavior, like all other activities of honey bees, results from their innate
> —and therefore unavoidable—responses to environmental stimuli. (Doull, 1971)

The opening sentence of Doull's interesting analysis of honey bee behavior is illustrative of a pervasive bias among biologists with respect to the relative importance of instinct and learning as regulators of insect behavior. This bias probably is the unavoidable and nondeliberate product of the paradigms (viewpoints from which nature is observed and interpreted) guiding practitioners within the major Western schools of animal behavior. As Kuhn (1962) points out, a paradigm defines "which problems [it is] more significant to have solved." The paradigm which unites a group of investigators also determines which experimental results and interpretations will be regarded as significant and important. They are those which strengthen the paradigm "by extending the knowledge of those facts that the paradigm displays as particularly revealing, by increasing the extent of the match between those facts and the paradigm's predictions, and by further articulation of the paradigm itself." Data and interpretations which do none of these tend to be rejected or ignored as being noncentral (peripheral) or irrelevant, and contribute little to the evolution of central theory and ruling dogmas.

It is probably fair to say that the major contemporary schools of animal behavior and comparative psychology have their roots in the nineteenth-century works of Darwin (1868, 1871, 1873), Romanes (1884), and Morgan (1896). By "roots," I mean both the establishment of a viewpoint from which animal behavior can be studied and the erection of dominant theories. The evolutionary paradigm established the expectation of hierarchical mental capabilities, with "lower" animals lacking the plasticity of behavior characteristic of "higher" forms such as man. Natural selection of adaptive responses would lead to development of fixed patterns of behavior. Thus the study of innate behavior emerged as a central thrust within the paradigm, and after the studies by the Peckhams (1898) and Fabre (1920) insects came to be regarded as essentially creatures of instinct. Subsequently, the emphasis on forced movements by Loeb (1918), on the "releasers" of programmed behavior by Lorenz (e.g., 1937), and on instinctual behavior by Tinbergen (e.g., 1951) and by their respective colleagues have served to strengthen and fix the paradigm.

It was within this framework of increasing emphasis on instinct and decreasing attention to plastic behavior that an understanding of honey bee sensory physiology and behavior developed. Nevertheless, early experiments in these areas incidentally provided data which should be interpreted in terms of associative conditioning and memory.

II. EXPERIMENTS ON BEHAVIOR OF BEES

Early experiments which evaluated honey bee perception of form (e.g., von Frisch, 1914; Hertz, 1929, 1931, 1933), color (e.g., Lubbock, 1875; von Frisch, 1914; Kuhn, 1927), and odors (e.g., von Frisch, 1919) had in common an implicit associative conditioning. Worker honey bees were "trained" to take sugar solutions from feeders associated with particular visual or olfactory signals. These training signals were then presented in the absence of food, either alone or in conjunction with alternative patterns, colors, or odors. Return of the "trained" foragers to the training signals provided evidence that the honey bees could perceive a signal and distinguish it from other competing environmental cues. Sensory abilities of honey bees, determined by experiments of this type, are summarized by Ribbands (1953) and von Frisch (1950, 1967).

More recent experimentation of this type has shown that honey bees have the capacity to differentiate the angle of lines or stripes on a vertical field (Wehner and Lindauer, 1966; Wehner, 1967), the forms of various four- and six-pointed stars (Schnetter, 1968), discs and stars of different diameters or areas (Weizsäcker, 1970), and flicker frequencies of a sinusoidally modulated light stimulus (Vogt, 1969). From these results, it is evident that conditioning of forager bees to associate food with a particular training stimulus is a necessary part of the experimental procedure. Nevertheless, analysis of experiments of this type usually has not been made *in terms of learning,* and there has been a tendency to infer that "stimuli—of which an example might be the specific shape and color of a flower that releases the approach of a bee—act on *innate releasing mechanisms* in much the same way as do releasers" (Klopfer and Hailman, 1967, p. 40, italics added).

One other area of honey bee reserach, peripheral to the direct study of learning, should be mentioned. This concerns recruitment of foragers to particular food sources in the field. As with the study of sensory abilities, experiments on forager recruitment usually involve the *training* of a group of foragers to a feeding station established in the field, followed by a study of arrivals of members of that group or of additional bees to that station. Early emergence of an "innate language" model for the recruitment mechanism (e.g., von Frisch, 1923, 1947; Lindauer, 1952) diverted attention from the learning element in such experiments. According to the "language" hypothesis, bees have evolved an instinctual symbolic communication system involving "dance" maneuvers on the part of successful foragers. Potential recruits are presumably programmed to instinctively interpret quantitative information encoded in the dance, and, upon release of foraging behavior by the dance, they supposedly fly directly to the food located at some specific site in the field.

The "dance language" hypothesis enjoyed favor for a considerable period of time, partly because it was consistent with the dominant "evolution of instinct" paradigm in animal behavior, and has been interpreted in those terms (e.g., von Frisch, 1967, Chapter 11). However, recent evidence inconsistent with predictions of that hypothesis (summarized in Wenner, 1971; Wells and Wenner, 1973) and an increased appreciation of the role of olfaction in forager recruitment (Wenner et al., 1969; Johnson and Wenner, 1970; Wells and Wenner, 1971) suggest that the entire body of data on recruitment might profitably be reinterpreted in terms of conditional, learned relationships between food sources and the odors associated with them. In this case, the "insects are creatures of instinct" paradigm would give way to one focusing on conditioned responses and plastic behavior.

In addition to the many investigations of sensory physiology and foraging behavior in which learning is an incidental component of design, there have been a number of explicit studies of learning phenomena in honey bees. Frings (1944) found that tethered bees could be trained to extend the proboscis in response to a conditional stimulus by pairing that stimulus with tarsal contact with a sugar solution. He used the conditioned response in a study of the honey bee olfactory sense organ. This technique also has been used to show that bees can be conditioned to extend the proboscis in response to water vapor or light (Kuwabara, 1957). After a conditioned proboscis response has been established, extinction will occur with presentation of a series (about ten) of nonreinforced conditional stimuli, with spontaneous recovery observable on the following day (Takeda, 1961).

Vareschi and Kaissling (1970) recently have repeated the conditioned proboscis response experiment using a greater variety of odors as conditional stimuli, and they have shown for the first time that drones as well as workers respond to conditioning. Queens also can be conditioned to olfactory stimuli (Vareschi, 1971). Furthermore, a considerable fraction of experimental animals showed the conditioned response after only a single rewarded exposure to the test odor. In one series of experiments, 30% of workers and 20% of drones which were conditioned showed single-trial conditioning (Vareschi, 1971). While most bees in Vareschi's experiments required only one or a few trials for conditioning, some individuals (25%) did not exhibit conditioned behavior after ten rewarded exposures to the training scent.

Conditioning experiments on unrestrained bees are of great interest to students of honey bee behavior, and results of such experiments may have implications for agriculture. Laboratory methods for such studies of learning phenomena were developed independently by Bermant and Gary (1966) and by Wenner and Johnson (1966). Wenner and Johnson used single-frame observation hives connected by one entrance to the outside and by another to a 1.2 by 0.3 m enclosed ramp leading into the laboratory (Fig. 1). Bees could fly freely from the hive at any time or could walk onto the observation ramp within the laboratory. The ramp was white, translucent, and lighted

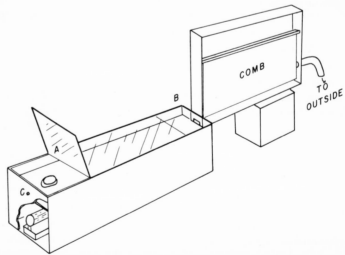

Fig. 1. An apparatus for conditioning honey bees in the laboratory. Bees enter the training ramp at hive entrance B. Sugar solution may be provided by lifting lid A or by introducing by pipette through hole C (from Wenner and Johnson, 1966).

from below. Five centimeters above the ramp was a glass lid with red filter (transmitted $\lambda > 580$ nm) such that investigators could observe the bees without disturbing their behavior. The distal 18 cm of the lid was hinged and could be opened for purposes of providing food or marking individual bees.

Bees were easily trained to walk the length of the ramp and to feed from a dish of 2 M sucrose solution. Whenever the dish was empty, most bees remained in the hive or near the entrance, However, when the food supply was replenished, they resumed regular visitation and feeding at the dish.

In a first series of experiments, opening' the lid (more precisely, the resultant air current) was a conditional stimulus. Unrewarded bees did not respond to that stimulus. However, when opening of the lid was followed by provision of sugar solution during a number of irregularly spaced trials, bees responded to the conditional stimulus by moving to the feeding area prior to provision of food. The Wenner and Johnson (1966) data provided evidence of conditioning after 20–25 rewarded trials presented at the rate of three trials per day. After 30 trials, more than 50 bees exhibited the conditioned response. Since this was a mass conditioning experiment, the exact number of trials required for conditioning of any individual bee was not determined.

Extinction and spontaneous recovery were readily demonstrated using this system. After a group of bees was conditioned by an appropriate series of rewarded trials, extinction was achieved by exposing the bees to three or four unrewarded trials during a day. After 1 day of extinction trials, the conditioned response was again evident on the following morning. However, after a second day of unrewarded trials no recovery occurred.

Using the same apparatus, Wenner and Johnson (1966) showed that light or odor cues, as well as air movements, could serve as conditional stimuli. Bees could readily be conditioned to move onto the ramp, and to the feeder location, whenever the light below the ramp was turned on. Similarly, after a series of rewarded exposures to a particular odor (in this case, smoke), bees would move rapidly to the feeder when that odor was blown into the hive.

One additional variation of the above experimental design is of interest. The feeding area was divided into two equal parts, one with a blue door and one with a yellow door. After presentation of an odor cue to which the bees were conditioned, food was always provided behind the yellow door. The positions of the blue and yellow cues were randomly varied during successive trials. Within four trials, 31 bees were conditioned to move directly to the yellow-doored chamber upon presentation of the odor cue in the hive. In this situation, the bees discriminate between cues of the same modality (visual cues). Bermant and Gary (1966) also demonstrated color discrimination conditioning in a similar apparatus and, in addition, incorporated reversal conditioning in their experimental program. In addition to discriminating among cues of the same modality, conditioned bees followed a succession of cues of different modalities presented in a particular order (Wenner and Johnson, 1966). The visual cue to which the bees were conditioned was present continuously. The animals responded to it, however, only after exposure to the conditional odor stimulus.

In addition to experiments discussed above, a considerable literature has resulted from studies of honey bee conditioning at the I. P. Pavlov Institute of Physiology, Leningrad, USSR. Bees have been conditioned to protrude the sting in response to a light signal after several simultaneous exposures to light and electric shock (the unconditional stimulus). Similarly, when light and elevated temperature were paired, bees could be conditioned to ventilate their hive by fanning in response to light alone (Lobashev, 1958).

Discrimination conditioning to different colors has been shown in maze studies requiring bees to turn left when presented with a green signal and right when given a blue signal, or the reverse (Nikitina, 1959a,b). Additional maze studies have shown that bees can be conditioned to follow a particular sequence of color signals to a food source (e.g., Lopatina and Chesnokova, 1959; Chesnokova, 1959). After conditioning, bees would walk to a feeder only when blue, green, and red color cues were presented in the training sequence, and only when the series was complete. Lopatina and Chesnokova (1962) also trained bees to follow a linear sequence of large painted color cues under field conditions and found that the conditioned response was retained when the hive was moved to a new location and the color signs were placed in a different field.

The importance for the economy of the honey bee colony of conditioned responses to odors associated with food and to remembered sequences of environmental stimuli merits additional emphasis. Dependence of bees on familiarity with a territory for their homing ability is well documented. Although experienced forager honey bees captured at the hive and released at a considerable distance from it can find their way home, bees unfamiliar with the landmarks or released in areas without landmarks fail to return to the hive (Romanes, 1885). In a more recent experiment, Wolf (1926) moved a hive to an unfamiliar site, captured and marked 18 bees before allowing any flight from the new location, and released them at distances ranging from 150 to 700 m. None of the bees returned to the hive. However, after 1–4 hr of flight activity in the new location, most bees tested could home from 50 m, and as the duration of free flight activity increased, the distance bees could be transported and still find their way home also increased.

Levchenko (1959) demonstrated a correlation between age of workers and homing ability. Bees 5 days old could not home from 100 m. At 10 days of age, however, 80% successfully homed from that distance but not from 200 m. At the age of 18 days, 57% returned to the hive after being transported 500 m. Homing ability was not retained when the hive was moved to a new location, indicating that familiarity with the vicinity rather than age was the important factor. When bees in a new location were rapidly trained to visit a feeder at 1500 m, they could home along the training path but not from any other direction. Becher (1958) also noted that young bees without flight experience rarely succeed in homing from 50–100 m, but after a single orientation flight they can home (success decreasing with distance) after displacements up to 700 m.

Thus learning and memory are essential to successful foraging by honey bees. Orientation flights, during which bees familiarize themselves with visual landmarks and characteristic odors of a territory, must precede food gathering. During foraging flights, memory of visual and olfactory landmarks is reinforced, and the foraging range may be expanded by exploration of territory more distant from the hive.

That bees have an effective long-term memory has been shown by Menzel (1968). Bees were conditioned to associate a spectral color (444 or 590 nm) with a food reward of 2 M sucrose. After three rewarded trials during which bees drank their fill, choice of the training color in discrimination tests was approximately 80%. This preference was not at all diminished after 2 weeks, indicating that duration of memory exceeds that interval. If rewards associated with a training stimulus were short (3 sec), bees learned as rapidly as with long rewards but memory decayed within 30 min. Thus phenomena similar to the short-term and long-term retentions known for vertebrates have been recorded for honey bees.

Reversal training is possible after conditioning to one of the above colors and requires more trials after longer periods of initial training. After four or five reversals in succession during a training period, bees reverted to random behavior and chose the two colors with equal frequency (Menzel, 1969), and only after a day away from the test situation could they again be conditioned to a color.

After a single food-rewarded exposure to a distinctive odor, 90% of bees distinguished that scent from others and from an unscented source (Koltermann, 1969). If, after a rewarded trial with one scent, bees were given a rewarded trial with a second scent, they chose the first in a discrimination test. However, after one rewarded trial with a first scent followed by five rewarded trials with the second one, only 18% chose the first training scent. But the first scent was not forgotten. When a discrimination test was performed using scent No. 1 and a third scent new to the experiment, 98% of the experimental bees chose the first training scent. Thus memory of a scent persisted even during an interval when bees showed a conditioned response to a different odor. Duration of long-term memory of odors was not determined, but conditioning to an odor showed little decay in 4 days (Koltermann, 1969).

The role of conditioned responses to odors in the mobilization of an experienced work force to its food sources has been examined by Johnson and Wenner (1966). About 40 marked bees were trained to regularly collect clove-scented 1.5 M sucrose from a dish placed 150 m from an observation hive. When the dish was allowed to remain empty, individuals occasionally inspected it. The cumulative number of different bees to inspect the dish was linear, with an average of 7.3 individuals making such inspections during eleven 15-min tallies (Fig. 2). However, when the dish was refilled after being empty for at least an hour, the cumulative number of different bees to arrive at the dish was exponential (Fig. 3), with nearly all of the trained bees arriving within 10 min. Since there was no dancing in the observation hive during the exponential phase of this experiment, the inference is inescapable that this work force showed a conditioned response to presence of the food odor on bodies of returning successful foragers. This would account for the exponential nature of the recruitment curve.

As a test of the conditioned response model, a second group of workers was trained to regularly feed from a dish of lavender-scented sucrose at a place 150 m from the hive in a direction opposite to that of the first (clove-scented) station. After both dishes were allowed to remain empty for a time, Johnson and Wenner (1966) could selectively direct either segment of the work force to visit its (empty) dish by blowing the appropriate scent into the hive. Furthermore, when the same training scent was used at both stations and the dishes were allowed to stand empty, the particular bees accustomed

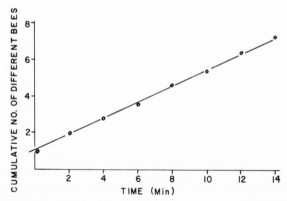

Fig. 2. Cumulative average for 11 tallies of marked bees monitoring an empty dish at which they had been previously successful (from Johnson and Wenner, 1966).

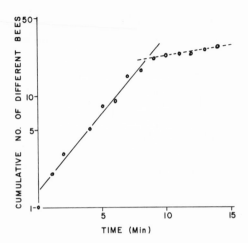

Fig. 3. Exponential increase (within the first 9 min) in the number of different bees visiting a dish newly refilled with scented sucrose solution (from Johnson and Wenner, 1966).

to feeding at each station were rerecruited to it when only one dish was refilled and only one bee allowed to carry the odor back into the hive (Johnson, 1967). These are clear demonstrations of the effectiveness of discrimination conditioning to odors for the mobilization of segments of the experienced work force of a hive to food sources in times of yield. Since nearly all field bees of a colony are attached to various crops at any given time, this mechanism maximizes the efficiency with which nectar and pollen are brought into the hive.

In addition to discrimination conditioning and conditioning to sequences of stimuli discussed above, honey bees may show instrumental conditioning. Nuñez (1970), for example, has developed an artificial nectary which will deliver a programmed quantity and rate of sugar solution whenever the experimental bee depresses a lever within it with her head. He has used this device effectively to study relationships between sugar flow and behavioral parameters, but our concern here is with the training of honey bees to operate an instrument. Each experimental bee was first allowed to feed at the artificial nectary without pushing the lever. After conditioning to that source, food was provided only when depression of the lever activated a delivery pump. Bees quickly mastered the task and would exert forces up to 300 mg for appropriate rewards. This is the more remarkable since the weights of individual worker honey bees are only 86.5 ± 3.8 mg ($\bar{X}$ and SD calculated from data in Wells and Giacchino, 1968). Pessotti (1972) also has shown discrimination and operant conditioning using honey bees and other bee species.

Honey bees show both Pavlovian conditional responses and Skinnerian operant conditional behavior; short- and long-term memory; conditioning to simple and complex stimuli; and reversal, extinction, and spontaneous recovery; thus they cannot be hierarchically separated from vertebrates on the basis of these qualitative learning phenomena. It is appropriate to ask, therefore, whether there exists experimental evidence of generalization, concept formation, or problem solving by honey bees. The reversion to random choice in multiple reversal experiments with colors noted by Menzel (discussed above) may be interpreted either as a generalization or as a form of extinction. Menzel (1969) further noted that bees successively trained to three colors would respond positively to all spectral colors tested. Also, generalization of training scents to similar odors which the bees can be conditioned to differentiate has been noted (e.g., Takeda, 1961). These studies, however, are not rigorously designed to provide answers to the question at hand.

Of greater usefulness are the data and interpretations of Mazochin-Porshnyakov (1969). He tested the hypothesis that bees can generalize and form concepts in a series of "extended discrimination" experiments. For instance, after conditioning bees to fly to a blue triangular figure in preference to an orange square one, he offered a test trial of orange triangular *vs.* blue square figures. Approximately equal numbers of bees chose each of the figures. If, however, training was to a white triangular figure on black background, tests showed preference for black triangulars on white to white squares on black. Furthermore, after training to distinguish an orange triangular figure from a blue square one, bees preferred green triangles to yellow squares. Thus bees appear to develop a concept of the triangular and, in the absence of confounding stimuli, display it in discrimination tests. As a further test, bees were conditioned to prefer triangles to squares of the same size, then offered choices between squares and triangles of greatly differing sizes. Triangles always were chosen, irrespective of absolute or relative size.

If we wish operationally to distinguish a concept from a percept, we need to establish a condition in which a behavior associated with a percept is evoked by a set of symbols which have the same relationship as those perceived during conditioning of that behavior but which have not previously been experienced by the test animal. By this operational definition, the conceptualization of triangle would seem to have been achieved by honey bees. By similar sets of experiments, Mazochin-Porshnyakov (1969) has operationally established the conceptualization of "multicolored and checkered." Bees conditioned to a checkered multicolored object such as a checkered pink and green rectangle preferred it to either a unicolor pink or a green rectangular signal of the same shape. Offered a test choice of blue, blue-green, or blue/blue-green checkered rectangular signals, bees chose the last. Choice of multicolored checkered over unicolored occurred for several color combinations tested. The response seems to have been generalized and conceptualized. In

the same paper, Mazochin-Porshnyakov reported other apparent concep-
tualizations such as "position of test object relative to other objects" and
inferred that "these investigations . . . lead to the conclusion, that it is per-
missible to speak of intelligent behavior in bees."

III. CONCLUSIONS

Whether or not we wish on present evidence to join Mazochin-Porshnyakov in his last inference, it should not be difficult to agree that honey
bees show a degree of plastic behavior unsuspected by those working within
the classic "insects are creatures of instinct" paradigm. Indeed, as Wenner
and Johnson in the United States, Kuwabara and Takeda in Japan, Menzel
and Koltermann in Germany, as well as Lopatina, Chesnokova, Lobashev,
and Mazochin-Porshnyakov in Russia have attempted experimentally to
display the extent and importance of learned behavior in bees, they have
operated from quite a different viewpoint. By casting off the restrictions of
the "instinct" paradigm, by accepting the possibility of significant plasticity
in honey bee behavior, and then by exploiting this capability, these investigators may have precipitated a small-scale Kuhnian revolution. It now seems
reasonable and constructive to view honey bees as exhibiting flexible, learned
behavior that is important to their ecology. Further experimentation within
the "learning" paradigm may tell us whether honey bees are capable of
insightful or problem-solving behavior. In any case, we are not justified in
regarding them as hierarchically distinct and inferior in learning abilities to
the majority of vertebrates. Honey bees do not appear to show more stereotyped behavior than grebes or sticklebacks nor learning phenomena quali-
tatively different from those exhibited by rats.

REFERENCES

Becher, L., 1958, Untersuchungen über das Heimfindevermögen der Bienen, *Z. Vergl.
Physiol., 41*, 1–25.
Bermant, G., and Gary, N. E., 1966, Discrimination training and reversal in groups of honey
bees, *Psychon. Sci., 5*, 179–180.
Chesnokova, E. G., 1959, Conditioned reflexes established in bees through chains of visual
stimuli, *Trud. Inst. Fiziol. Pavlova, 8*, 214–220, in Russian.
Darwin, C., 1868, "Variation of Animals and Plants Under Domestication," D. Appleton,
New York.
Darwin, C., 1871, "The Descent of Man," D. Appleton, New York.
Darwin, C., 1873, "Expression of the Emotions in Man and Animals," D. Appleton, New
York.
Doull, K. M., 1971, An analysis of bee behavior as it relates to pollination, *Am. Bee J.,
111*. 266.

Fabre, J. H., 1920, "The Wonders of Instinct," George Allen and Unwin, London.

Frings, H., 1944, The loci of olfactory end-organs in the honey bee, *J. Exptl. Zool.*, *97*, 123–134.

Hertz, M., 1929, Die Organisation des optischen Feldes bei der Biene. I, *Z. Vergl. Physiol.*, *8*, 693–748.

Hertz, M., 1931, Die Organisation des optischen Feldes bei der Biene. III, *Z. Vergl. Physiol.*, *14*, 629–674.

Hertz, M., 1933, Über figurale Intensitäten und Qualitäten in der optischen Wahrenehmung der Biene, *Biol. Zbl.*, *53*, 10–40.

Johnson, D. L., 1967, Communication among honey bees with field experience, *Anim Behav.*, *14*, 487–492.

Johnson, D. L., and Wenner, A. M., 1966, A relationship between conditioning and communication in honey bees, *Anim. Behav.*, *14*, 261–265.

Johnson, D. L., and Wenner, A. M., 1970, Recruitment efficiency in honey bees: Studies on the role of olfaction, *J. Apic. Res.*, *9*, 13–18.

Klopfer, P. H., and Hailman, J. P., 1967, "An Introduction to Animal Behavior," Prentice-Hall, Englewood Cliffs, N. J.

Koltermann, R., 1969, Lern- und Vergessensprozesse bei der Honigbiene aufgezeigt anhand von Duftdressuren, *Z. Vergl. Physiol.*, *63*, 310–344.

Kuhn, A., 1927, Über den Farbensinn der Bienen, *Z. Vergl. Physiol.*, *5*, 762–800.

Kuhn, T. S., 1962, "The Structure of Scientific Revolutions," University of Chicago Press, Chicago.

Kuwabara, M., 1957, Bildung des bedingten Reflexes von Pavlovs Typus bei der Honigbiene, *J. Fac. Sci. Hokkaido Univ.* (*Zool.*), *13*, 458–464.

Levchenko, I. A., 1959, Homing ability of bees, *Pchelovodstvo*, *36*, 38–40, in Russian.

Lindauer, M., 1952, Ein Beitrag zur Frage der Arbeitsteilung im Bienenstaat, *Z. Vergl. Physiol.*, *34*, 299–345.

Lobashev, M. E., 1958, Study of the instincts of the honeybee by means of conditioned reflexes, *Pchelovodstvo*, *35*, 21–25, in Russian.

Loeb, J., 1918, "Forced Movements, Tropisms, and Animal Conduct," Lippincott, Philadelphia.

Lopatina, N. G., and Chesnokova, E. G., 1959, Conditioned reflexes of bees to complex stimuli, *Pchelovodstvo*, *36*, 35–38, in Russian.

Lopatina, N. G., and Chesnokova, E. G., 1962, Motive stereotype of conditioned reflexes in the foraging activities of the honeybee, *Trud. Inst. Fiziol. Pavlova*, *10*, 245–254.

Lorenz, K. Z., 1937, The companion in the bird's world, *Auk*, *54*, 245–273.

Lubbock, J., 1875, Observations on bees and wasps. II, *J. Linn. Soc.* (*Zool.*), *12*, 226–251.

Mazochin-Porshnyakov, G. A., 1969, Die Fähigkeit der Bienen, visuelle Reize zu generalisieren, *Z. Vergl. Physiol.*, *65*, 15–28.

Menzel, R., 1968, Des Gedächtnis der Honigbiene für Spektralfarben, I. Kurzzeitiges und langzeitiges Behalten, *Z. Vergl. Physiol.*, *60*, 82–102.

Menzel, R., 1969, Des Gedächtnis der Honigbiene für Spektralfarben. II. Umlernen und Mehrfachlernen, *Z. Vergl. Physiol.*, *63*, 290–309.

Morgan, L., 1896, "Habit and Instinct," Edward Arnold, London.

Nikitina, I. A., 1959a, Conditioned signal reflexes in honeybees, *Nauch. Soobsch. Inst. Fiziol. Pavlov*, No. 1, 55–57, in Russian.

Nikitina, I. A., 1959b, Training the mobility of nervous processes in the honeybee, *Trud. Inst. Fiziol. Pavlova*, *8*, 157–164, in Russian.

Nuñez, J. A., 1970, The relationship between sugar flow and foraging and recruiting behavior of honey bees (*Apis mellifera*, L.), *Anim. Behav.*, *18*, 527–538.

Peckham, G. W., and Peckham, E. G., 1898, On the instincts and habits of the solitary wasps, *Wisc. Geol. Nat. Hist. Surv. Bull.*, No. 2.

Pessotti, I., 1972, Discrimination with light stimuli and a lever-pressing response in *Melipona rufiventris, J. Apic. Res., 11,* 89–93.

Ribbands, C. R., 1953, "The Behavior and Social Life of Honeybees," Bee Research Association, London.

Romanes, G. J., 1884, "Mental Evolution in Animals," Keegan, Paul, Trench, and Co., London.

Romanes., G. J., 1885, Homing faculty of Hymenoptera, *Nature, 32,* 630.

Schnetter, B., 1968, Visuelle Formunterscheidung der Honigbiene in Bereich von Vier- und Sechsstrahlsternen, *Z. Vergl. Physiol., 59,* 90–109.

Takeda, K., 1961, Classical conditioned response in the honey bee, *J. Insect Physiol., 6,* 168–179.

Tinbergen, N., 1951, "The Study of Instinct," Oxford University Press, New York.

Vareschi, E., 1971, Duftunterscheidung bei der Honigbiene—Einzell-Albeitungen und Verhaltensreaktionen, *Z. Vergl. Physiol., 75,* 143–173.

Vareschi, E., and Kaissling, K. E., 1970, Dressur von Bienenarbeiterinnen und Drohnen auf Pheromone und andere Duftstösse, *Z. Vergl. Physiol., 66,* 22–26.

Vogt, P., 1969, Dressur von Sammelbienen auf sinusförming moduliertes Flimmerlicht, *Z. Vergl. Physiol., 63,* 182–203.

Wehner, R., 1967, Pattern recognition in bees, *Nature, 215,* 1244–1248.

Wehner, R., and Lindauer, M., 1966, Die optische Orientierung der Honigbiene (*Apis mellifica*) nach der Winkerlichtungfrontal gebotener Streifenmuster, *Zool. Anz. Suppl., 30,* 239–246.

von Frisch, K., 1914, Der Farbensinn und Formensinn der Biene, *Zool. Jb. (Physiol.), 35,* 1–188.

von Frisch, K., 1919, Über den Geruchsinn der Bienen und seine blütenbiologische Bedeutung, *Zool. Jb., 37,* 1–238.

von Frisch, K., 1923, Über die "Sprache" der Bienen, *Zool. Jb. (Physiol.), 40,* 1–186.

von Frisch, K., 1947, The dance of the honey bee, *Bull. Anim. Behav., 5,* 1–32.

von Frisch, K., 1950, "Bees, Their Vision, Chemical Senses and Language," Cornell University Press, Ithaca, N. Y.

von Frisch, K., 1967, "The Dance Language and Orientation of Bees," Harvard University Press, Cambridge, Mass.

von Weizsäcker, E., 1970, Dressurversuche zum Formensehen der Bienen, insbesondere unter wechselnden Helligkeitsbedingungen, *Z. Vergl. Physiol., 69,* 296–310.

Wells, P. H., and Giacchino, J., 1968, Relationship between the volume and the sugar concentration of loads carried by honeybees, *J. Apic. Res., 7,* 77–82.

Wells, P H., and Wenner, A. M., 1971, The influence of food scent on behavior of foraging honey bees, *Physiol. Zoöl., 44,* 191–209.

Wells, P. H., and Wenner, A. M., 1973, Do honey bees have a language? *Nature, 141,* 171–175.

Wenner, A. M., 1971, "The Bee Language Controversy," Educational Program Improvement Corp,. Boulder, Colo.

Wenner, A. M., and Johnson, D. L., 1966, Simple conditioning in honey bees, *Anim. Behav., 14,* 149–155.

Wenner, A. M., Wells, P. H., and Johnson, D. L., 1969, Honey bee recruitment to food sources: Olfaction or language? *Science, 164,* 84–86.

Wolf, E., 1926, Über das Heimkehrvermögen der Bienen. I, *Z. Vergl. Physiol., 3,* 615–691.

Chapter 10

LEARNING IN GASTROPOD MOLLUSKS[1]

A. O. D. WILLOWS

Department of Zoology
University of Washington
Seattle, Washington, USA

I. INTRODUCTION

Among the six classes of mollusks, apart from that on the cephalopods the greatest amount of work relating to behavioral and neuronal plasticity has been done on the gastropods. The Gastropoda are a very large (over 80,000 species) and diverse class which have invaded every type of habitat with the exception of the aerial one. Their particular attractions to workers in neurophysiology and behavior are basically two. First, their diversity increases the likelihood that there will be one or several species readily available virtually anywhere on earth. A stroll in the marine intertidal is almost certain to turn up several snails, limpets, or nudibranchs, and numerous species of the families Lymnaeidae, Planorbidae, and Physidae inhabit pond, stream, and lake shores. There are snails and slugs inhabiting most leafy areas and dry grasslands. Even the sands of parched deserts support a gastropod fauna; e.g., *Sphincterochila* survives in the Negev (Yom-Tov, 1970; Schmidt-Nielsen *et al.*, 1971).

A second advantage from the point of view of the scientist interested in the neural correlates of behavioral phenomena is the extraordinarily large

[1] I wish to thank Dr. Peter Getting, Dr. Richard M. Lee, and particularly Dr. Alan J. Kohn for reading carefully and criticizing the first draft of the manuscript. Also, I am grateful to Eleanor H. Ridgway, who made the drawings for Figs. 3, 4, and 5. Stuart Thompson and Dr. Peter Getting assisted with some of the original experiments reported here. Research was supported by National Science Foundation Research Grant GB33720X to the author.

size of the neuron cell bodies in the ganglia of many of these creatures. Not all large gastropods have large neurons. Many of the largest neurons occur in the opisthobranchs and a few pulmonates, but there seems to be no rule by which one can predict size. For instance, many large marine prosobranch snails such as *Polinices* (the moon snail) have relatively small neuron somas (less then 50 μ). Terrestrial giants such as *Archachatina* and *Strophocheilus* have a few large neurons (200–400 μ) (Nisbet, 1961; Amoroso *et al.*, 1964; Faugier, personal communication). But some enormous neurons, perhaps the largest in the animal kindgom, are found in the medium and large marine opisthobranchs such as *Aplysia* and *Tritonia* (Fig. 1). These cells range up to a millimeter in diameter and sometimes can be seen with the naked eye. They are electrically excitable and often have yellow, orange, or red intracellular pigments. Between these extremes are a great many species having numerous cell bodies whose diameters span the range 50–500 μ. The additional fact that most of the very large neurons are similar in size, position, and pigmentation in different individuals and are therefore reidentifiable in different animals of the same species is of great value to the experimentalist, since he can return to the same cell from time to time in the same animal or in different animals in order to repeat experiments or to test ideas that have been generated by earlier work.

Size is also of great importance from the point of view of the neuro-chemist interested in behavior since the contents of single neurons can sometimes be analyzed after dissecting the neuron free. If something is known of the electrophysiological or behavioral function of a particular cell, then correlations can be sought with components of the cytoplasmic or nuclear contents. Significant progress in these directions has already been made. Transmitter substances and enzymes involved in their metabolism have been studied in gastropod ganglia and neurons by Giller and Schwartz (1968) and McCaman and Dewhurst (1970). Kuhlmann (1969), Lasek and Dower (1971), and Coggeshall *et al.* (1970) have studied the nuclear DNA content of single neurons.

Fig. 1. A, The sea hare, *Aplysia californica*. An opisthobranch mollusk with numerous large and identifiable neurons in its ganglia, this common Pacificspecies has been studied extensively in both behavioral and neurophysiological experiments. Typically, animals range from 10 to 20 cm in length and weigh 100–500 g (specimen photographed in Sea of Cortez by E. Janss, Jr.). B, The brain of *Tritonia tetraquetra*. In this and many other gastropod species, large neuron cell bodies distributed over the surfaces of the ganglia naturally contain white, yellow, orange, red, or black pigments. Such neurons have been identified by their sizes, locations, and pigmentation, and in some cases their specific roles in the control of behavioral responses have been determined. The brain is 7 mm across, and neurons having diameters from 50 to 1000 μ are typical.

Fig. 1A

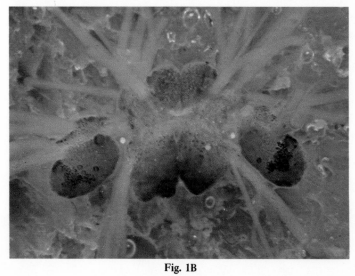

Fig. 1B

In addition to the interest engendered by their diversity as an animal group and the extraordinarily large size of the nerve cell bodies in their nervous systems, the gastropods are often well suited to behavioral experiments, since, taken as a group, their repertoire of activities is extremely diverse. In addition to the usual sorts of simple reflexive responses to noxious or other stimuli, there are examples of complex feeding, locomotory escape, and mating behaviors that are well adapted to neurobehavioral study.

II. SYSTEMATIC ORGANIZATION

A. Phylogenetic Relations

There are six classes in the phylum Mollusca (Fig. 2), of which four are larger and more commonly known: the Amphineura, which include chitons and primitive ancestral forms; the Gastropoda, e.g., limpets, snails, and slugs; the Bivalvia (=lamellibranchs =pelecypods), which include clams, oysters, mussels, and scallops; and the Cephalopoda, e.g., octopus and squid. There are differing views on the divergent evolution of the molluscan classes from a common ancestor. The most commonly accepted one is that all classes evolved from a common ancestor which is now extinct. Another proposal is that the Amphineura or Monoplacophora may represent ancestral forms to the other classes. One feature that unites all classes is a larval

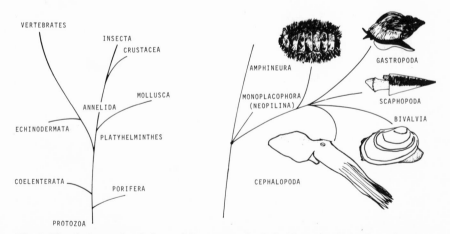

Fig. 2. The phylogenetic relations of the gastropods. Left: The animal kingdom, showing major animal groups and their evolutionary affinities. Right: The classes of mollusks. The phylum may have arisen from a segmented ancestor, phylogenetically related to the annelid worms. The several classes of mollusks may have descended from a common ancestor. The Amphineura and *Neopilina* (Monoplacophora) have primitive features that suggest placing them on the evolutionary line of descent of the other classes.

development consisting of trochophore-like and free-living veliger stages (Fig. 3) which precede settling and major metamorphosis into a crawling or swimming young adult. Excepting the Monoplacophora, the gastropods are all unsegmented and have a small coelomic cavity, a well-developed hemocoel that is often important for locomotion as a hydrostatic skeleton, and a muscular foot. The radula, further discussed below, is a unique feeding structure in the molluscs. Most other anatomical features are variable through the phylum.

The gastropod class is conveniently divisible into three subclasses (the classification of all species mentioned in this chapter is provided in Table I): (1) the Prosobranchia (mostly marine snails and limpets, e.g., *Littorina, Buccinum,* and *Acmaea*), in which all the adults show a torted (pleurovisceral nerve loop twisted in a figure 8) nervous system and the gills and anus are usually in a mantle cavity at the anterior region of the body; (2) the Opisthobranchia, which are mostly attractively colored marine slugs (e.g., *Navanax, Doris,* and *Aplysia*) and in which torsion is usually apparent in the veliger larva and partly or completely lost in the adult; and (3) the Pulmonata (mostly terrestrial and freshwater snails and slugs, e.g., *Lymnaea, Physa, Helix,* and *Limax*), which have also lost the torted nervous system form in the adult but by an apparent shortening of the nerves of the pleurovisceral connective loop to the point that most ganglia have fused in the cephalic region.

A further division (Morton, 1967) of the three subclasses helps to establish a frame of reference for their classification. The prosobranchs are usually put into the following three orders: The Archaegastropoda, including mostly less specialized marine forms, are virtually all herbivorous and usually have top-shaped spiral shells or are flattened and have no coiling, as in the limpets of the families Patellidae and Acmeidae. The Mesogastropoda are

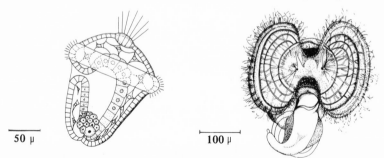

50 μ   100 μ

Fig. 3. Gastropod larval forms. The trochophore (left) has a ciliary band around its midsection and an apical tuft. Generally, in gastropods the trochophore stage is passed over within the egg. In aquatic forms, the veliger (right) is usually released into the plankton and uses the water currents produced by its twin-hooded ciliary fringe for both feeding and locomotion. After a short time, the veliger undergoes a major metamorphosis into a small version of the adult.

Table I

Class Gastropoda		
Subclass	Prosobranchia	
Order	Archaegastropoda	(a)
Order	Mesogastropoda	(b)
Order	Neogastropoda	(c)
Subclass	Opisthobranchia	
Order	Cephalaspidea	(d)
Order	Anaspidea	(e)
Order	Thecosomata	(f)
Order	Gymnosomata	(g)
Order	Sacoglossa	(h)
Order	Acochilidiacea	(i)
Order	Notaspidea	(j)
Order	Nudibranchea	(k)
Subclass	Pulmonata	
Order	Basommatophora	(l)
Order	Stylommatophora	(m)

Acmaea	(a)		*Littorina*	(b)
Acteon	(d)		*Lottia*	(a)
Adalaria	(k)		*Lymnaea*	(l)
Agriolimax	(m)		*Melibe*	(k)
Akera	(d)		*Melongena*	(c)
Ampullaria	(b)		*Nassarius*	(c)
Aplysia	(e)		*Natica*	(b)
Archachatina	(m)		*Navanax*	(d)
Archidoris	(k)		*Notarchus*	(e)
Arion	(m)		*Nucella*	(c)
Buccinum	(c)		*Oliva*	(c)
Bullia	(c)		*Onchidium*	(m)
Calliostoma	(a)		*Paludina*	(b)
Calyptraea	(b)		*Patella*	(a)
Capulus	(b)		*Philine*	(d)
Chromodoris	(k)		*Physa*	(l)
Conus	(c)		*Planorbarius*	(l)
Crepidula	(b)		*Planorbis*	(l)
Dendronotus	(k)		*Pleurobranchaea*	(j)
Diodora	(a)		*Plocamphorus*	(k)
Doris	(k)		*Polinices*	(b)
Elysia	(h)		*Rumina*	(m)
Facelina	(k)		*Siphonaria*	(l)
Fasciolaria	(c)		*Sphincterochila*	(m)
Fusitriton	(b)		*Stagnicola*	(l)
Gibbula	(a)		*Strombus*	(b)
Haliotis	(a)		*Strophocheilus*	(m)
Helisoma	(l)		*Tegula*	(a)
Helix	(m)		*Terebellum*	(b)
Hermissenda	(k)		*Thais*	(c)
Ianthina	(b)		*Tritonia*	(k)
Latia	(m)		*Trivia*	(b)
Limacina	(f)		*Trochus*	(a)
Limax	(m)			

a diverse order having representatives in most marine environments and are both herbivorous and carnivorous in feeding habits. The common shore winkle *Littorina* is an example of a common mesogastropod. Typically, their shells are taller spirals, and many have an anterior shell extension which projects the inlet siphon and sensory osphardium forward, but there are limpet-like examples including the Chinamen's hat and slipper limpets as well (*Calyptraea* and *Crepidula,* respectively.) The Neogastropoda are all carnivorous or carrion feeders, and most, like the Mesogastropoda, have elongate, spiral shells with an anterior shell canal. Common examples include *Buccinum, Conus, Thais,* and *Oliva.*

The opisthobranchs, although once divided into two large groups, the tectibranchs and nudibranchs, are now most commonly placed into eight orders (for classification of several common examples, see Table I).

Pulmonates are grouped into two orders according to the location of the eyes. Both groups have two pairs of tentacles on their head. The Basommatophora have eyes at the bases of the posterior pair of tentacles, while the eyes of Stylommatophora are at the tips of these tentacles. The former group includes *Lymnaea* and *Planorbis,* while the latter includes the snail *Helix* and the slug *Agriolimax.*

B. Evolution of Nervous Systems and Anatomy GASTROPODS (?)

To put the gastropods into perspective in regard to their behavioral capabilities, their gross anatomical characteristics, and their diverse and

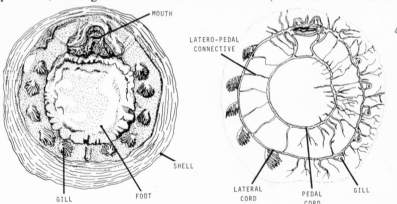

Fig. 4. *Neopilina galathea* (left, seen from the ventral aspect), the only known living genus of the Monoplacophora, considered by some to link the mollusks to a segmented ancestor. Its external gills and internal structures are segmentally arranged. The nervous system (right) has serially arranged cross-connections between pedal and pleural cords, with approximately one such connective per gill. Nerve cords are medullary, as in the chitons and primitive gastropods, with no prominent ganglion-like swellings.

sometimes peculiar neuroanatomy, it is helpful to remind oneself of the evolutionary relationships of the class.

To begin with, the mollusks are probably derived from a segmented ancestor. Until, 1957, the most compelling evidence for this was the close phylogenetic relationship that apparently exists between mollusks and the segmented annelid worms. Cleavage patterns, the determinacy of the developing cells, and even the early trochophore larval stages of both phyla bear a strong resemblance to one another. But a segmented body plan was not known in any living mollusk until 1957, when Lemche reported his dramatic find of a metameric single-shelled mollusk among the material from a deep dredging expedition made 7 years earlier on the Danish ship *Galathea* (Fig. 4). Even in this animal (*Neopilina galatheae*), however, there is no obvious external segmentation except possibly the gills. It is the internal anatomy, especially that of the nervous system, that has identical repeated structures. Although nerve commissures are not evident except in the head region, there are exactly ten pairs of connectives[2] extending between the two pairs of longitudinal nerve trunks.

The tendency toward segmentation which persists and is far more fully developed as the dominant form of body organization in the arthropods declines very rapidly through the classes of mollusks. In the gastropods, three trends are dominant, and all of these counter the metameric ancestral influence (Fig. 5). First, the orderly pairing of nerve trunks disappears altogether. This is particularly true of commissures between the pleural cords and of connectives between pleural and pedal nerves. There are few direct lateral interconnections between the longitudinal cords posterior to the head region in gastropods. An exception to this is that regular commissures pass from one pedal connective to the other in primitive forms and are modified into an irregular plexus of cross-connections in the foot of most gastropods.

Second, the nerve cell bodies formerly distributed in the longitudinal trunks become aggregated into a few pairs of nodelike ganglia along the cords. These swellings along the pleural nerves become the visceral, intestinal, and parietal ganglia of the pleurovisceral loop. The neuron somas of the pedal cords are withdrawn, figuratively speaking, to the pedal ganglia in the head region, and the pedal cords then disappear entirely, except in the archaegastropodan prosobranchs. Finally, the ganglia themselves become increasingly concentrated in the region of the head and in some cases achieve a level of organization and specialization characteristic of a "brain." Increas-

[2]Connectives are distinguished from commissures on the basis that connectives interconnect dissimilar ganglia on one side of the midline, while commissures join pairs of analogous ganglia across the midline. Thus the nerve directly connecting, for instance, the left pedal to the left pleural ganglion is a connective, while the nerve connecting one cerebral ganglion to the other cerebral ganglion is a commissure.

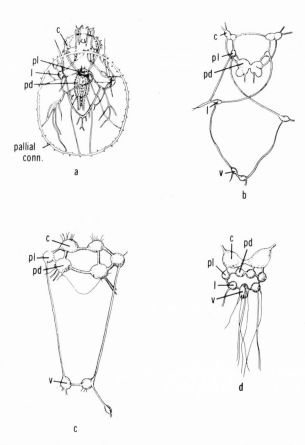

Fig. 5. Evolutionary developments in gastropod neuroanatomy. Three trends are evident in these examples from (a) an archaegastropodan prosobranch, *Megathura,* (b) the mesogastropod *Pomatia,* (c) the opisthobranch *Aplysia* (in *Aplysia californica,* not shown, the parietovisceral ganglia are fused by loss of the commissure), and (d) the pulmonate *Lymnaea.* Medullary nerve cords become ganglionated, torsion is lost, and ganglia become concentrated in the head region by shortening of connectives as we pass through this sequence. Abbreviations: c, cerebral; pl, pleural; I, intestinal; pd, pedal; v, visceral.

ing cephalization occurs with the evolutionary development through the prosobranchs and is a common feature of opisthobranchs (e.g., *Aplysia* and *Tritonia*). It is especially prominent in pulmonates. In *Helix,* for instance, all ganglia are arranged in a periesophageal ring, with the cerebrals placed dorsally. Each of these latter is composed of three enlargements, of which one, the procerebral lobe, is considered by some to be specialized as an association area.

Looking briefly to the forms that have evolved in parallel with the gastropods, it is quite evident that these same trends continue there, too (Morton,

1967). The scaphopods, an infrequently studied group of bilaterally symmetrical creatures living in tusk-shaped shells open at both ends, have prominent ganglia and well-developed cephalization. The bivalves, lacking a
head, show little evidence of concentration of ganglia in the anterior region
of the body, although in most the pleural and buccal ganglia are fused with
the cerebrals. Concentration of ganglia does occur, however, in the visceral
region of some species, e.g., the scallop *Pecten* (Dakin, 1910), where visceral
and parietal ganglia are fused together and form a multilobed nervous center
with inputs from osphradia, various pallial nerves, and the cerebral ganglion.
In the cephalopods, tendencies to cephalization and fusion of ganglia are
fully expressed in a highly developed brain.

C. Torsion

In all gastropod species, with the exception of a few primitive prosobranchs, the nervous system anatomy can be understood most simply as a
circumenteric ring of ganglia including the cerebral, pleural, and pedal ganglia
surrounding the esophagus with a second pleurovisceral connective loop
attached posteriorly. Along this loop are usually located other ganglion pairs
including parietal, intestinal, and visceral (abdominal).

Torsion primarily affects the pleurovisceral loop (Fig. 5). In the primitive untorted condition, the loop is in a plane extending posteriorly and
dorsally from the pleural ganglia. To form a correct impression of the torted
condition, one can imagine that the caudal half of the loop has been twisted
in a counterclockwise sense (seen from the dorsal aspect) through nearly
180°, while the anterior half remains fixed. Thus the original visceral ganglion
at the posterior extremity of the loop has exchanged sides and is turned
ventral side dorsally. The original left and right intestinal ganglia that were
somewhere near the halfway point along the loop are now below and above
the visceral mass, respectively. This latter fact accounts for their being re-
named sub- and supraintestinal ganglia, respectively, in some species.

The anatomical picture is modified in many gastropods by the tendency
for the ganglia of the pleurovisceral loop to migrate along the cords toward
the head in an asymmetrical fashion. For instance, in *Aplysia* it is supposed
that both intestinal ganglia, the visceral ganglion, and the *right* parietal
ganglion have fused to form the pleurovisceral ganglion, while the *left*
parietal ganglion has migrated to the left pleural ganglion. This interpretation
accounts for the location of left and right giant cells (originally in left and
right parietal ganglia, respectively) in the left pleural and pleurovisceral
ganglia, respectively (Hughes, 1967).

In summary, in its overall evolutionary development, the central nervous
system anatomy of gastropods can be described as passing through three

stages. The most primitive is that which occurs in ancestral forms, and in both *Neopilina* and various Amphineura. The nervous system consists of poorly defined ganglia dispersed along two pairs of parallel nerve cords. The pedal ganglia or medullary cords are interconnected at intervals by commissures. At the next level, that of many prosobranchs, nerve cell bodies become incorporated into ganglia and the above-described torsion takes place. Most connectives and commissures are lost. The third and most highly developed condition is that of the opisthobranchs and pulmonates, where torsion has been largely undone and a considerable degree of cephalization of ganglia has taken place.

III. DEVELOPMENT

Little attention has been paid to the likelihood that gastropod learning, like that of other species, depends on the developmental stage of the animal. Neither have attempts to modify gastropod behavior in genetic experiments been reported. In part, both these limitations arise out of the time required and the difficulties encountered in rearing gastropods in the laboratory. As neurophysiological and behavioral studies continue to involve gastropods, however, it is clear that knowledge of developmental aspects and techniques of laboratory rearing will be increasingly useful and relevant.

A. Copulation and Egg Laying

The reproductive behavior of gastropods has been carefully studied in species from all three major groups but never in the context of "learning" or its underlying neurophysiology. An extraordinarily broad range of behavioral patterns is involved in copulation, egg placement, and brooding.

Prosobranchs differ markedly from the opisthobranchs and pulmonates in their reproductive habits. Separate sexes are typical in the prosobranchs, and cross-fertilization is the rule. The exceptions to the rule include a few mesogastropod species in which parthenogenetic development occurs and a small number of limpets and snails that cross-fertilize but are hermaphrodites. Prosobranch hermaphroditism tends to be "consecutive" and protandrous; i.e., the animal is first a functional male and later in life, usually after copulation, is transformed into a female. In some species, several reversals of sex occur. It is obvious that such transformations require profound changes in anatomy and behavior.

On the other hand, the opisthobranchs and pulmonates are essentially all hermaphroditic. Most are "simultaneous" hermaphrodites; i.e., both eggs and sperm mature together in time. Eggs and sperm are generated alongside one another in a common ovotestis and are usually separated to prevent self-fertilization either by a thin physical barrier as in *Helix arbustorum* (Buresch, 1912, reported in Fretter and Graham, 1964), for instance, or by the timing of the production and release of male and female gametes. Autosterility adds to the barrier against self-fertilization in some genera, notably *Helix* and *Agriolimax* (Fretter and Graham, 1954). However, numerous examples, especially those whose habits in life reduce the likelihood of encountering other of the species, are quite capable of self-fertilization and do so either by simply permitting eggs and sperm to mingle during their passage down the common genital duct or by engaging in self-copulation. In this latter instance, e.g., *Lymnaea* spp, and *Planorbis planorbis*, fertilization results from the animal thrusting its penis into its own vaginal duct and releasing sperm there (Colton, 1918; Crabb, 1927).

Copulatory behavior during cross-fertilization in the gonochoristic (separate sexes) and hermaphroditic species is extremely varied (see, e.g., Chase, 1953, aerial mating in slugs; Fretter and Graham, 1964, for review). A few examples will serve to illustrate this diversity.

In *Helix*, shortly after initial contact, genital apertures are externalized and both animals nearly synchronously propel a calcareous dart from the female opening toward the other animal. Sometimes this dart is thrust with sufficient violence to break through the skin of the other animal and may even pass deep into the organs of the body cavity. Whether this bizarre behavior is an example of "sadomasochism in snails" as suggested by Hutchinson (1930) or is a form of stimulation to ensure species recognition, its occurrence is followed shortly afterward by coupling of the two animals, each transmitting sperm into the other. A similar behavior, involving a quite different anatomical mechanism, has been reported for *Onchidium* by Webb *et al.* (1969). In opisthobranchs, partial detorsion has occurred, bringing the anus and genitalia to the right-hand side of the body wall. Coupling is affected after a period of circling during which appropriate postural changes are made, genital apparatus are enlarged, and frequent body contact is made. Usually, the hermaphroditic animals then simply align themselves head to tail, right side to right side, and insert their respective penises into the vaginal opening of the other. In *Aplysia* and *Lymnaea stagnalis* as well as other hermaphroditic species, numerous animals may copulate simultaneously, forming lengthy chains in which each animal acts as a male to the next one ahead or below (Barraud, 1957). Although courtship and coupling may last much longer, exchange of sperm in all these cases normally continues only a few minutes.

Sperm is transmitted either free or in semisealed packets (spermatophores). In either case, it is then stored in a receptacle on a sidebranch of the genital duct and awaits ripening of eggs. This may be weeks or even months away. The most likely stimulus for spawning in the normal environment is

a seasonal fluctuation (usually a rise) of a few degrees in temperature. Go-
nadal maturation including oocyte development in *Aplysia punctata* can be
stimulated in the laboratory by maintaining the temperature at 15°C, which
is 5–8°C higher than field temperatures (Smith and Carefoot, 1967). From
the point of view of laboratory rearing, factors associated with capture and
artificial maintenance (perhaps merely the close physical contact of numerous individuals) facilitate copulation, and accordingly egg laying is a frequent
event in the laboratory with pulmonates (e.g., *Limax*) opisthobranchs (e.g.,
Aplysia, Hermissenda, Tritonia, and *Archidoris*), and prosobranchs (e.g.,
Buccinum and *Fusitriton*).

Most often, eggs are encapsulated in a shell or albumin or both and are
then deposited individually or in groups. Terrestrial snails place their eggs in
small depressions in the substrate or in buried locations. Marine and freshwater species deposit massive ribbons of capsules containing 10^2 to 10^6 eggs.
MacGinitie (1934), for instance, records that one *Aplysia* laid nearly one-half
billion eggs within a period of just over 4 months. Emerging strings of eggs
are engulfed in a mucous ribbon and grasped by the marginal foot region
and by this means firmly attached to the substrate. The detailed appearance
and structure of the ribbon and the pattern (spiral, simple loop, or draped)
of its placement are extremely varied and species specific. Once the eggs
have been placed, the parent usually leaves them and pays no further attention to their condition (e.g., *Helix, Aplysia, Tritonia,* and *Lymnaea*). Brooding of egg capsules takes several forms (see Fretter and Graham, 1962). The
prosobranch *Fusitriton* attaches its egg capsules in a compact mass on the
rocks, then remains on top of them until the veligers emerge (C. Eaton,
personal communication). The family Calyptraeidae (e.g., *Capulus* and
Crepidula) similarly remain over their egg masses and irrigate the developing
embryos in their inhalent water flow (Ankel, 1936; Lebour, 1937; Thorson,
1946). Other prosobranchs (e.g., the cowrie *Trivia*) insert egg capsules into
chambers and then seal them into the body walls of their normal food
animals (often tunicates) or attach them to the outer surface. A few species
bear the eggs within the oviduct and release young veligers (e.g., *Ianthina
janthina*).

B. Larval Stages

Beween fertilization and the appearance of the adult gastropod, the
animal passes through two stages (Fig. 3). The first, the trochophore, bears
a close resemblance to the comparable stage in annelid and platyhelminth
development. In all but the most primitive archaegastropod species, it is
passed over within the egg. The trochophore is a bilaterally symmetrical
creature with a ciliated band encircling its midregion. The second stage is the

veliger, and it is during this stage that torsion occurs in all mollusks. In most gastropods, torsion occurs as a distinct event in the veliger stage involving contraction of an asymmetrically placed muscle band. In opisthobranchs, the evidence suggests instead that the organs of veligers *arise* in the torted condition developmentally and furthermore that the extent of torsion is less (Thompson, 1962; Tardy, 1970). This latter fact may account for the apparent partial detorsion in the adults. The veliger resides in a wide-pitch spiral, calcareous shell and bears a double hood, fringed with cilia used for locomotion and to create a water current from which food may be caught. The time at which this feeding begins depends on the time of release from the egg, which in turn depends on the amount of nutritive materials in the form of yolk or so-called albumin in the egg. Some species have exceedingly yolky eggs which sustain the larvae until quite late in the veliger stage. Others having fewer nutritive stores in the egg are released earlier and depend heavily on planktonic feeding for sustenance. Once released, the free-swimming veliger ultimately settles to the bottom and a major metamorphosis occurs, transforming it into a small version of the adult. Development through trochophore and veliger stages occurs completely within the egg in many pulmonates, small snails or slugs emerging directly from the egg capsule in these cases.

The problems encountered in laboratory rearing through the larval stages are usually in regard to providing the appropriate food. The free-swimming aquatic forms sometimes require particular plankton and other as yet unspecifiable environmental conditions for their maintenance to stimulate the final metamorphosis from veliger to small adult. Scheltema (1962), Fretter and Montgomery (1968), and Pilkington and Fretter (1970) have shown that veligers of several prosobranchs including *Nassarius* and *Crepidula* grow optimally only under controlled conditions of light, temperature, and food supply (both abundance and species of food are important). Success in maintaining *Strombus* through metamorphosis has been reported by D'Asaro (1965). The veligers of *Archidoris, Adalaria,* and *Tritonia hombergi* have been reared to metamorphosis (Allen and Nelson, 1911; Thompson, 1968; Thompson, 1962). Many years of successful culture of commercially valuable bivalves have shown that trial and error may be required to determine nutritive preferences and that food size and species are very important (Walne, 1964). The veliger must be able to engulf the prey, and nannoplankton having one dimension of less than 10 μ are likely to be most suitable.

The pulmonates and modern prosobranchs that undergo development through trochophore and veliger stages within the egg are perhaps the most immediately feasible species for laboratory rearing. No special attention need be paid the developing embryos other than to provide the most rudimentary conditions of temperature, humidity, and protection from predators and decay bacteria.

Curiously, the most difficult species to rear past metamorphosis have been those marine opisthobranchs having large, yolky eggs (Thompson, personal communication). These veligers remain within the egg and obtain their nutrition from the included yolk store until quite late. Despite a successful escape from the egg, and with only a brief period remaining before time of normal metamorphosis, they have thus far proven difficult to induce beyond the free-swimming stage.

An important environmental factor in the induction of metamorphosis in some species having yolky eggs has been found to be an adequate supply of the normal adult food. For instance, Thompson (1962) has raised the British nudibranch *Tritonia hombergi* past metamorphosis to adulthood by providing a source of the anthozoan *Alcyonaria* (dead-man's fingers) at the time the veliger settles. Applying this knowledge to a congener with larger eggs has not been successful, however. *Tritonia diomedia,* for instance, has not yet been coaxed past the veliger stage in the laboratory, despite the presence of supplies of its normal adult food, the sea whips, *Virgularia* and *Ptilosarcus* (Thompson, personal communication). This difficulty suggests that more subtle environmental factors are involved in the induction of metamorphosis in these cases.

IV. GASTROPOD BEHAVIOR

Virtually all of the motor activities of gastropods occur very slowly by vertebrate or arthropod standards. This means that the initial impression gained by an observer of normal attention span and patience may be that gastropods do very little of behavioral interest. Partly to counter this widely held impression and further to establish a picture of the normal behavioral repertoire of gastropods, against which past and future learning experiments can be considered, a description of certain of the sensory capabilities and common behavioral patterns of the class is worthwhile here.

Virtually all animal behavior depends on stimulation for its initiation or its coordination or both. Accordingly, to put the behavior of gastropods into context, it is appropriate to review briefly their sensory capabilities. Such a review is especially important here because the gastropods are primarily responsive to a different class of stimuli from those important to most other animals, including the arthropods, vertebrates, and even the cephalopods. While most other members of the animal kingdom are keenly dependent on visual inputs, the gastropods have at best only rudimentary form vision, little or no capability for color discrimination, and only primitive ability to determine relative distances and movement. Their world is principally a chemical and tactile one, and their receptors and behavior reflect this difference.

A. Chemical and Tactile Senses

While receptors for chemicals and touch are probably distributed diffusely over most of the epithelium, these senses are especially keen in particular areas. The cerebral or oral tentacles, the rhinophores, the leading edge of the foot, the area around the mouth (the lips), and the siphon are all known to be particularly sensitive to both forms of sensory input (Charles, 1966; Kohn, 1961). These receptor areas are used individually or in concert (1) to locate food at a distance (homing on food from several meters in calm water and crawling upstream in a current towards food as much as a meter away by prosobranchs have been reported by Kohn, 1961), (2) to discriminate between prey organisms based on contact chemoreception (e.g., *Navanax* feeding behavior; Paine, 1963), (3) to trigger escape responses in prosobranchs and pulmonates (see below), (4) in reproductive behavior to find mates in gonochoristic species (e.g., *Paludina viviparus;* see Wolper, 1950) and to sense secretions that trigger mating in hermaphroditic species (e.g., *Chromodoris;* see Crozier and Arey, 1919), (5) in homing to follow olfactory cues (e.g., *Helix pomatia;* Edelstam and Palmer, 1950), and (6) in the locomotor response to reimmersion in seawater after a period of desiccation caused under normal circumstances by a low tide (e.g., *Patella;* Arnold, 1957).

Prosobranchs, most opisthobranchs, and the aquatic pulmonates have an additional chemosensory or tactile sensory organ located in the mantle water stream near its entrance. This structure, the osphradium, has been implicated in testing the quality of the incoming water stream for nutrients, dissolved gases, or toxic agents and also as a tactile receptor for the detection of suspended silt in the respiratory water stream (Yonge, 1947; Kohn, 1961).

The behavioral and neurophysiological data bearing on the specific function of the osphradium are compatible with either a chemosensory or an osmoreceptor function (Copeland, 1918; Brown and Noble, 1960; Michelson, 1960). Bailey and Laverack (1963, 1966) working with *Buccinum* and Jahan-Parvar *et al.* (1969) with *Aplysia* have found that osphradial stimulation with food or hypo- or hypertonic salt solutions elicited marked central neural responses. In *Aplysia,* the responses were noted principally in white neurons of the parietovisceral ganglion and appeared to be initiated a long way from the cell body. Action potentials may have been generated in the osphradium itself. The on-going firing of neuron R15 was abruptly inhibited by exposure of the osphradium to hypoosmotic solutions, i.e., below isotonicity with seawater (Stinnakre and Tauc, 1966, 1969). Other agents such as seawater recently containing seaweed (*Aplysia's* food), amino acids, high P_{CO_2} or P_{O_2}, or low P_{O_2} had no effect on the activity of the neuron. It is possible, of course, that other central neurons received inputs from receptors for these and other stimuli affecting the osphradium. Obviously, the function of the osphradium is not limited by this finding.

The osphradium is developed as a long ridge containing many cilia and sensory cells and is well innervated by axonal processes from the osphradial ganglion in species that live in heavily silted locations, e.g., *Buccinum,* and inconspicuous in others that live in cleaner water. For instance, in *Aplysia* the osphradium is merely a small patch of sensory epithelium at the anterior base of the gill, close to the union of the efferent brachial vein with the anterior gill margins. The difference seems also to be well correlated with feeding habit, the carnivorous prosobranchs having prominent osphradia while the herbivorous opisthobranchs do not.

Szal (1971) reports the discovery of a new chemoreceptive organ at the base of each gill leaflet in archaegastropods. On the basis of behavioral experiments, its function is supposed to be linked to detection of approaching starfish. The approach of starfish such as *Pisaster* causes *Tegula,* for instance, to climb rapidly out of the water.

Abundant structural evidence points to the presence of receptor neuron terminals for touch and chemoreception free in the epithelium, and the associated cell bodies are usually found in or directly beneath the epithelium. The endings themselves are not specialized or elaborate and consist of single or branched naked terminals. In most cases, the terminals converge into a short connecting process which leads through the epithelium to the cell body below. A single axon is directed centrally and presumably transmits conducted action potentials to the central nervous system.

Studies on mechanoreceptors located in the epithelium of the mantle shelf of *Aplysia* (Kandel, personal communication) point to a central location in the parietovisceral ganglion for the associated receptor cell bodies. Impulses recorded in certain pleurovisceral ganglion cells synchronously follow brief, local mechanical perturbations of the skin up to 30/sec. Increasing intracellular hyperpolarizations of the soma cause the impulses generated by mechanical stimulation to be blocked farther and farther from the soma, presumably out the axon toward the sensory terminals. There is no sign of an excitatory postsynaptic potential on the leading edge of the spike, or in isolation after the spike has been blocked by hyperpolarization, which might be expected if a primary receptor neuron precedes the neuron in question.

Laverack and Bailey (1963) studied the activity produced in peripheral nerves by stretch-sensitive terminals near the base of the siphon in the whelk *Buccinum,* and although they found three receptor types ("on" response, "on–off" response, and "on–off" response superimposed on spontaneous firing with slow adaptation) their evidence does not suggest a location for the receptor cell bodies, i.e., central or peripheral.

Taken together, the available structural and electrophysiological evidence points to a peripheral location for most receptor somas. At the same time, there are almost certainly some receptor terminals whose somas are located in central ganglia. The apparent division of receptor neurons into classes having centrally and peripherally located somas deserves further attention. It is becoming increasingly important that details of sensory to motor pathways be determined, including the location of cell bodies and synapses, since in many cases it is these cell bodies and synapses that hold the key to separating sensory adaptation from centrally mediated phenomena in learning experiments.

B. Geotaxis

The main organ of geotaxis in gastropods is the statocyst. It is a partially transparent, spherical body 100–500 μ in diameter. The statocyst is usually at or near the union of cerebral–pedal or cerebral–pleural ganglia. Study of the histology of these organs (e.g., in *Helix*) suggests a rather simple mechanism for their function. The inner surface of the spherical enclosure is lined with flattened neuronal somas, 60–90 μ in largest diameter, that have numerous cilia embedded in their surfaces. The cilia are directed into the statolymph-filled cavity. Loose in the cavity are many small concretions (statoconia) that are free to bump against the cilia. Presumably, the animal determines its geotactic orientation by the combination of activated neurons in the statocyst walls. In some species, the statoconia are highly reflective and appear to be in constant vibratory motion in living preparations, suggesting that the cilia are motile.

Wolff (1969, 1970) recorded from the whole static nerve and from individual fibers of the nerve while rotating the pulmonates *Arion, Limax,* and *Helix* about all principal axes. He concluded that there are about 12 hair cells and that they all respond in a phasic–tonic fashion to rotations about all axes. However, each receptor cell appears to respond maximally over a slightly different range. Furthermore, they are strictly gravity detectors with peak response, measured as the summed frequency of all the units together, occurring when the animal is turned upside down, and they do not respond selectively to rotational accelerations.

C. Vision

Most gastropods have a pair of eyes located in the region of the head. They may be on the end of mobile and retractable eyestalks, e.g., *Agriolimax,* or *Terebellum,* or closely attached to the surface of the central ganglia, with no well-defined optic nerve, e.g., *Melibe*. Eyes are reduced or completely lost in a number of burrowing species. An unusual exception to this tendency are the well-developed eyes and eyestalks used as periscopes in the burrowing strombid *Terebellum* (Abbott, 1962). The primary structures range from simple pigment cups to lens-eyes with simple corneas to a few more complex eyes having retinas with moderately densely packed receptor cells (10,000 receptors per square millimeter for octopus; Young, 1962). Some have the potential for limited accommodation over a narrow range (Newell, 1965).

The nerve cell bodies of the various ganglia are often densely pigmented, and the possibility has been considered (Charles, 1966) that some degree of direct light detection may take place at the ganglia themselves. This idea gained support from studies of Arvanitaki and Chalazonitis (1960, 1961) that showed that monochromatic light could elicit "on," "on–off," and "off" responses in some cells and complete inhibition of spiking activity in others. However, the light intensities required to produce these responses were quite high with respect to the intensities expected at the ganglia in nature. The ganglion used (the parietovisceral ganglion of *Aplysia*) would normally be buried beneath several millimeters of tissues, including a densely pigmented, dark epithelium. It seems unlikely that the incident light levels normally encountered by the animals would be sufficient to act directly on ganglion cells in a fashion usable as vision.

Another, perhaps more likely function for these pigments as electron acceptor substitutes for molecular oxygen has been suggested by Zs.-Nagy (1971). He finds that the relative density and prominence of the lipochrome pigments judged subjectively in different bivalve and gastropod species are related to the tolerance of the species of anoxibiotic stress. The species having most pigments can withstand the longest period of anoxia. More convincingly, however, the redox potential of an anoxic physiological solution to which slices of the pigmented molluscan nerve tissue (*Lymnaea* and *Anodonta*) have been added is reduced more slowly over a longer period of time than an identical solution to which, for instance, frog brain slices have been added. This suggests that a component (presumably the lipochrome) of the molluskan brain slice is acting as an electron acceptor in the electron transport chain, in the place of the absent normal acceptor, oxygen. A similar suggestion assigning a metabolic role for the yellow carotenoid pigments of *Lymnaea* is made by Karnaukhov (1971), except in this case the author proposes that the pigment acts as an auxiliary respiratory pigment (along with myoglobin) to bind and store molecular oxygen for ultimate use in the cytochrome chain.

Virtually nothing is known of the visual pigments of gastropods, and very little electrophysiological work has been done. Barth (1964) and Dennis (1967) have recorded intracellularly from the large (30–50 μ) photoreceptor neurons of *Hermissenda crassicornis*. They showed that on-going spiking activity, which appeared to originate at some distance from the soma, could be either excited or inhibited (depending on the cell) by photic stimulation. Barth's conclusion that the observed inhibition was probably due to inputs from other cells and not primary was borne out by Dennis's work. This is in contrast to the light-generated inhibition of *Aplysia* parietovisceral ganglion cells and that shown for the photoreceptors of the pallial nerves of *Spisula*, a bivalve, by Kennedy (1960). Dennis found that the eye contained only five photoreceptor cells and that each of these was connected to all others by an inhibitory pathway. Thus the observed "on," "on-off," and "off" responses could be explained in terms of the interactions between photodepolarization and synaptically mediated inhibition. An additional point that is of special interest for behavioral work is that the individual photoreceptor units had different receptive fields, indicating the possibility that the animal may be able to discriminate photic inputs coming from different locations with respect to the body axis.

Although the gastropod photosensory apparatus is often quite simple and in some cases reduced or missing, there are a few behavioral responses which are well defined and useful as subjects of experimental study in the context of learning. One, the withdrawal response of *Lymnaea* to shading, will be discussed in detail in a later section. Another is the withdrawal of the anal branchial plume of *Chromodoris zebra* (Crozier and Arey, 1919). A shadow cast over the strongly photopositive animal elicits withdrawal of the branchial plumes within 0.6–1.0 sec. Although the animal has eyes located beneath the skin behind the rhinophores, they are apparently not involved in the gill withdrawal reflex. For withdrawal to occur, a shadow must fall specifically on the gills themselves, and the response occurs in isolated preparations of the gills. The photoreceptors that initiate this response are as yet undescribed.

Light or its absence exerts a marked kinetic effect on many gastropods. Again in *Chromodoris* (Crozier and Arey, 1919), shading of the whole body invariably elicits a reduction in locomotor activity. If offered a choice between two halves of a shallow aquarium, one side intensely lit (e.g., direct sunlight) and the other side well shaded, most animals will be found in the *shade*. This curious finding serves to emphasize both the positive phototropism and the dominant photokinesis of the animal, for although it is attracted to the brightly lit side, immediately upon arrival there its wandering movements increase in velocity dramatically. The resulting accelerated locomotion inevitably brings the animal back to the shaded area, where its movements once again slow down. Since its rate of movement is greatly reduced in the shaded area, it spends more time there finding its way back to the bright side. Hence, on average, animals are found predominantly on the dark side.

Although *Helix* is negatively phototropic, nevertheless it will move away from dark patches on a light background toward the more brightly lit regions, suggesting that the dark patches are recognized as being unattractive for some reason other than the poor luminance (von Buddenbrock, 1919). The prosobranch *Littorina* is also negatively phototropic (Fraenkel, 1927). However, this tropism is reversed when the creature is crawling upside down underwater. The adaptive value of this reversal may be that it helps to increase the likelihood that the animal will follow the descending tide out of crevices and caves in the rocks. There are examples of prosobranchs (e.g., *Littorina*), opisthobranchs (e.g., *Elysia*), and pulmonates that show rather complex light compass reactions (menotaxes). Burdon-Jones and Charles (1958) showed that *Littorina littoralis,* for instance, orients on the shore with respect to the sun, although it may reverse its direction of movement 180° from time to time.

D. Locomotion and Feeding Behavior

To date, virtually all studies of plasticity in gastropod behavior have centered around withdrawal or other defensive reflexes. By their nature, these acts are of very short duration and are often ballistic in the sense that once triggered by an appropriate stimulus they are executed to completion without any further dependence on sensory input. These kinds of responses have so far resisted all regimens designed to induce learning, with the exception that habituation often occurs.

Locomotion and feeding behaviors, on the other hand, are typically composed of much more prolonged sequences of activity that depend on frequent or continuous reference back to sensory cues originating in the environment. In part because of these differences, such activity will probably prove to be of greater potential interest for workers who are concerned with operant and classical conditioning.

1. Pedal Locomotion

Under normal conditions, the vast majority of gastropods search for food and mates and escape from noxious stimuli by gliding upon an uninterrupted sequence of wavelike contractions that pass over the foot. A mucous trail laid down by secretions from gland cells in the foot increases the grip on the substrate. The basic mechanism of this process has been studied by Lissman (1945a,b) and is shown in Fig. 6. In its simplest form, bands of the foot are lifted up, moved forward, and then replaced on the substrate. There are two closely coupled sets of muscular contractions to consider. Narrow transverse bands of vertical or oblique dorsal–ventral muscles are responsible for lifting the foot and then permitting it to return to the substrate. The main longitudinal muscle sheets also contract in very short bands extending across

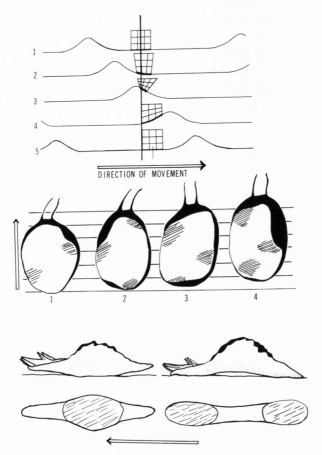

DIRECTION OF MOVEMENT

Fig. 6. Locomotion on the gastropod foot. Top: Contraction waves moving from posterior to anterior on the foot shown at five successive points in time. The waves sequentially lift and then replace transverse bands extending across the foot. The anterior portion of the lifted band is shortened, thereby moving that local foot region forward. The posterior portion of the lifted band is expanded before being replaced slightly ahead of its original position. Shortening caused by muscles in the vertical and longitudinal planes at each of the five stages is indicated by relative lengths of lines of box. Many such waves occupy the foot at any one time. Middle: Ditaxic locomotion in *Haliotis,* the abalone (after Lissman, 1945*a*). Numerous waves of contraction, essentially like those described above, occupying limited areas (shaded) on each half of the foot are shown at four successive points in time. Areas occupied by contraction waves pass over the foot separately on the left and right halves from posterior to anterior, successively moving alternating portions of the food forward. Bottom: Locomotion in *Aplysia.* Initially (left), the anterior foot is protracted and then appressed again to the substrate. The portion of the foot making contact with the substrate then moves posteriorly along the foot (right). When the area of contact reaches the middle part of the body, the anterior part protracts again, taking another "step."

the foot, resulting in horizontal displacement in the anterior–posterior direction. Forward body movement can be visualized in its rudimentary form as beginning with the lifting of a narrow band a few millimeters wide extending across the foot at its trailing edge. Another contraction wave in the longitudinal muscles occupying the anterior half of the lifted band results in relative movement along the substrate. The foot is not invaded by one wave at a time, however. Instead, there are many (separated by 1–5 mm) moving anteriorly along the foot at any moment. This is the normal mechanism of movement on the foot in *Helix, Polinices,* and many dorids, eolids, and pulmonates (for reviews, see Clark, 1964, and Morton, 1967). A different mechanism, not involving contraction in longitudinal muscles and instead using only dorsal–ventral contractions and a hydraulic system of blood spaces, has been proposed for the limpet *Patella* by Jones and Trueman (1970).

The locomotor mechanism of the abalone *Haliotis* depends on anteriorly directed contraction waves as above, but these waves occur in patches on both sides of the midline of the foot and the whole foot is not simultaneously invaded (Fig. 6). A roughly circular patch occupying perhaps one-fourth of the area on one side of the foot develops these waves beginning in the tail region, and then the circular invasion patch progresses along the foot toward the head. As this patch moves forward, a second begins to form at the tail on the other side of the midline and the two patches then progress forward, the one leading the other by one-fourth to one-half the length of the foot. As the first reaches the anterior foot margin, a new patch begins to form on the same side at the posterior margin, and so on. The effect resembles bipedal perambulation in a sack.

Other variations include examples in which the wave progression is from anterior to posterior (i.e., retrograde) and others in which the waves travel along the foot following indirect paths over part of the distance transversely or splitting into two halves and then rejoining. Waves may move across the foot diagonally (Miller, 1969).

Two other quite different but important locomotor mechanisms are sometimes observed. One, recorded in *Aplysia* (Parker, 1917), involves a sort of galloping movement in which each "pace" is equal to about one-fourth of the animal's length. The foot is normally long and slender, and during this kind of movement it is made even narrower. The leading edge is raised up off the substrate and then protracted upward and ahead and then placed back down. The arch thus formed extends over as much as one-third of the animal's body length. The raised arch progresses caudally, and as it reaches the tail, the leading edge is again raised to take another "step." This sort of movement is used as an alternative to the more common pedal waves described above under circumstances when speed is required (e.g., *Polinices* and

Helix). There are a number of prosobranchs and opisthobranchs in which contractile waves are never seen and still others in which arrhythmic muscular waves are typical. In species that do not locomote by pedal muscular waves, forward movement is achieved by the activity of huge numbers of backward-beating cilia which are distributed all over the foot. The ciliary stroke is made against the highly viscous resistance of the mucous path laid down ahead.

2. Leaping, Twisting, and Swimming

In some species, particularly prosobranch snails in the family Strombacea, normal locomotion is accomplished as a series of quick leaps or jumps (Parker, 1922; Yonge, 1937). The response appears fully integrated and co-ordinated at 3 weeks past metamorphosis upon the first, presentation of the stimulus (Berg, 1971). The animal first raises and protracts the leading edge of the foot and the curved opercular shell, then replaces them on the substrate, gaining purchase in the substrate by digging the operculum in slightly. Then the visceral mass and shell are brought along as well with a vigorous twitchlike contraction of the longitudinal muscles of the column. Each leap advances the animal as much as one-half its body length.

In numerous prosobranchs, including *Haliotis, Nassarius,* and *Buccinum,* contact with starfish or other echinoderms elicits a violent escape reaction. The shell is rapidly twisted through a 180° arc, beginning with a 90° turn in one direction and followed by 90° in the opposite. The sequence is repeated several times or until the stimulus is removed. The movement is surprisingly violent and is usually effective in breaking any initial tube-foot contact that may have been made by the potential predator. The twisting response is normally accompanied by accelerated movement on the foot, with larger and more rapid waves along the foot or a sudden switch from ciliary or smooth-wave locomotion to the "galloping" described above. Similar responses have been reported also for several species of the limpet *Acmaea* (Margolin, 1964; Bullock, 1953). Swimming is used as an escape in numerous species and in others for "place-to-place" locomotion of the sort that brings the animal into contact with food and mates (Fig. 7). The neuromuscular apparatus that controls these movements varies widely. Generally, swimming consists of one of the following patterns (for review, see Farmer, 1970): (1) a prolonged series of side-to-side flexions, each of which bends the head 45°–90° with respect to the main body axis, e.g., in the nudibranchs *Dendronotus* (Agersborg, 1922) and *Melibe* (Agersborg, 1923; Hurst, 1968); (2) an extended series of ventral–dorsal flexions in the sagittal plane again involving the main body wall musculature, e.g., in *Tritonia* (Willows, 1967, 1969) and *Hexabranchus* (Morton, 1964; Edmunds, 1968); (3) recurrent waves passing along or flapping movements of the muscular parapodia,

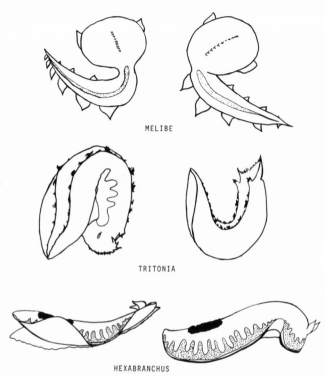

MELIBE

TRITONIA

HEXABRANCHUS

Fig. 7. Swimming behaviors in opisthobranch mollusks. In *Melibe* (top), swimming occurs both spontaneously and in response to noxious stimulation and consists of a prolonged sequence of lateral flexions. The body is bent alternately left and right by contractions in longitudinal muscles of the lateral body wall. There is no apparent control over the direction of swimming, and the animal is seen swimming in all orientations with respect to the substrate. In *Tritonia* (middle), swimming is elicited normally in response to skin contact with certain starfish. As in *Melibe*, control over the orientation of the animal with respect to the substrate seems not to exist. Propulsion is achieved by a shorter sequence (three to ten cycles) of dorsal and ventral flexions. The lateral margins of the body wall are flapped up and down in *Hexabranchus* (bottom) simultaneously with dorsal and ventral flexions similar to those seen in *Tritonia*.

e.g., in *Aplysia* (Parker, 1917) and *Akera* (Morton and Holme, 1955). *Hexabranchus emarginata* combines both movements (2) and (3) during swimming. In a few species, particularly the teropods such as *Limacina* (Morton, 1954), virtually permanent pelagic activity is typical, and swimming is accomplished by oarlike flapping or sculling motion of the parapodia.

Escape responses other than simple withdrawal and swimming are very common (Feder and Christensen, 1966; Ansell, 1969). These include secretion of purple or black fluids (e.g., *Aplysia* and *Acteon*), secretion of acid fluid (e.g., *Philine* and *Fusitriton*), jet-like swimming (*Notarchus*), autotomy of body parts (e.g., *Helix* and *Hermissenda*), production of luminous flashes (e.g., *Plocamphorus*), discharge of luminous secretion (e.g., *Latia*), and burrowing into the substrate (e.g., *Planorbis* and *Terebellum*). Other very common responses include extension and waving of mantle or cephalic tentacles (e.g., *Haliotis* and *Patella*), extension of the mantle lobes over the shell (e.g., *Polinices* and *Diodora*), raising of the shell by elongation of the columellar muscle (e.g., *Lottia* and *Acmaea*), turning away from the stimulus (e.g., *Calliostoma* and *Tegula*), and movement upward on vertical surfaces, to get out of the water (e.g., *Trochus* and *Gibbula*).

3. Feeding Behavior

Both the mechanics of feeding and the preferences in gastropod molluscs (see Owen, 1966, for review; also Thompson, 1964; Laxton, 1971) have become increasingly relevant for those studying either neurophysiological or behavioral aspects of gastropod learning. On the one hand, food detection and homing toward food require numerous and sometimes quite subtle discriminations of features of the environment. These discriminations are potentially modifiable. As well, food can be used effectively as a reinforcer in experiments to shape the behavior of an animal or to condition its responses. As was the case with the sensory capabilities of these animals, we also find that there are important differences between the mechanisms used by gastropods for feeding and those used by all other major animal groups. Accordingly, a brief description of the details of these unusual mechanisms is required.

All gastropods have a unique feeding apparatus (Fig. 8). The mouth and the esophagus are conventional and function essentially as they do in other animals. However, the buccal mass, which is connected to the mouth via an oral tube and transmits the food to the esophagus, has several features which are to be seen nowhere else in the animal kingdom. It is made up of three well-defined structures: the jaws (not present in all species), the odontophore, and the radula. The odontophore arises from the floor of the buccal cavity near the back and is stoutly muscular and mobile. The radula is a chitinous, flexible sheet covered by numerous rows (symmetrical about the midline) of recurved teeth which are directed dorsally and posteriorly, i.e., toward the esophagus. It is attached as a sheath to the exposed surfaces of the odontophore, but the odontophore is free to slide back and forth beneath it. The radular teeth are continuously replaced as they are worn off by the abrasive materials encountered during feeding. The radular sheet is free to slide over the odontophore and does so whenever the stiffened odontophore protracts or retracts. The odontophore–radula combination is used cyclically both to rasp and to bite off food bits. Rasping is usually accomplished by first pro-

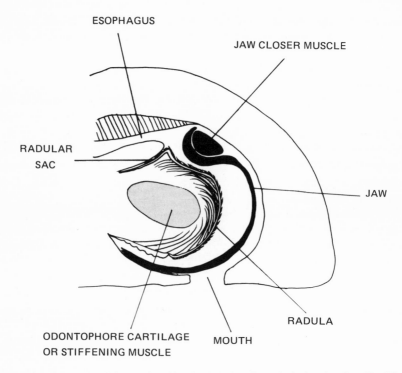

ESOPHAGUS

JAW CLOSER MUSCLE

RADULAR
SAC

JAW

RADULA

ODONTOPHORE CARTILAGE   MOUTH
OR STIFFENING MUSCLE

Fig. 8. The gastropod radula seen in midsagittal section through the head region. Used for biting or rasping food, the radula–odontophore complex is a unique adaptation of the mollusks. See text for description of functions.

tracting the odontophore–radula complex, appressing it to the substrate, and then moving the odontophore backward. Simultaneously, the radula slides forward and upward over the surface of the odontophore. The principal rasping action is at the point where the radula is deformed by the tip of the odontophore, since at this point the radular teeth temporarily change their orientation from downward and forward to upward and backward (Fig. 8). At this point, very small bits of food are broken off the substrate and subsequently carried as on a tooth conveyer belt upward and backward into the buccal cavity toward the esophagus.

The odontophore–radula combination can be used to nip off larger food chunks, too, independently of the jaws (e.g., in *Aplysia, Pleurobranchaea,* and *Tritonia*). To understand the mechanism of radular biting, one must recognize that a central longitudinal strip of the odontophore musculature can be withdrawn actively and collapsed inward, thereby bringing the lateral aspects of the radula into contact with one another. In species that employ this form of biting, the lateral teeth are longer and stronger than the central ones, and they act to pierce and hold onto the food so that it can be torn off the substrate.

Chitinous jaws, usually hinged at the top only and opening with a vertical aperture, are found in those numerous species that must bite off large chunks of food (e.g., leaves, algae, or sponge). The jaws are located at the point where the oral tube opens into the buccal cavity, and they are both opened and closed actively by prominent muscles that are attached on the outside surfaces of the jaws at both top and bottom. The jaws are used to bite off larger chunks of food in the more conventional way. Both types of biting, i.e., radular and that involving the jaws, are sometimes used together. This cooperative use can be illustrated by describing the feeding sequence of *Aplysia punctata* (Howells, 1942; Frings and Frings, 1965). Initially, the whole buccal mass is moved forward by means of both extrinsic and intrinsic muscle sheets attached to the body wall and buccal mass. The mouth is then opened, the jaws are opened, and the odontophore and radula are brought in contact with the weed which the animal is about to eat. At this time, the central region of the odontophore is collapsed and the radular teeth grip the weed. Simultaneously with and following this radular "bite," the radula is pulled around the end of the odontophore and the combination of pinching and rotating movements pulls the bit of weed a few millimeters into the buccal cavity in the direction of the esophagus. Then, without complete closing of the mouth or jaws, the cycle repeats and the radula is once again brought downward and forward and regrabs the weed further forward. This cycle is repeated until as much as 2 or 3 cm of seaweed is dragged into the mouth. At this point, midway through a cycle, the jaws are tightly closed around the weed and the piece that is in the mouth is torn from the remaining weed.

Biting with the radula or with the jaws or with both is quite common among opisthobranchs. The prosobranchs use the radula more commonly for abrading, rasping, or scooping of food particles. In *Littorina,* for instance, the radula is pressed against the substrate and then the odontophore is moved backward across it several times in sequence. At each passage, the radular teeth are raised and then lowered again, thereby abrading the substrate. The path followed by the odontophore across the radula changes slightly from pass to pass and thereby several different regions are successively abraded before the radula is withdrawn, carrying the combed out or loosened bits of food with it into the buccal cavity. Alternately, the odontophore may make but a single passage across the radula between each partial withdrawal. Each cycle is executed over a position slightly to one side of the previous one. After a sequence of six or ten such scrapes in a row, the animal moves forward and continues the rasping cycles. The progress of such a gastropod as it grazes across a patch of food can be seen as a series of scratch marks which proceed first to the left, then slightly forward, and then to the right, and so on. During this type of feeding, the radula and odontophore are only occasionally withdrawn completely into the mouth and apparently the food which is scraped free is continuously carried into the esophageal portion of the buccal mass by the cyclic scraping movements of the odontophore–radula. As opposed to this type of feeding, where the radula itself is not moved and the odontophore slides across it, other species including *Lymnaea* and *Patella* slide the whole radula with erect teeth across the substrate, moving both the odontophore and the radula together in a forward direction.

Numerous specializations of the radula and the odontophore occur, permitting several strange and unusual feeding behaviors. For instance, in the genus *Conus* the radular teeth are used as projectiles and to transmit salivary poison into the prey. Other species such as *Natica* and *Fusitriton* employ the radula and odontophore together with potent salivary secretions to drill holes in the shells of bivalve mollusks and echinoderms. There are examples of aquatic gastropods that feed on suspended matter. For these groups, including *Crepidula,* the slipper limpet (Yonge, 1938), the opisthobranch *Melibe* (Hurst, 1968), and certain species of the pulmonate pond snail *Lymnaea* (Morton, 1967), radular feeding is less important, and in some, e.g., *Melibe,* the radula is lost altogether. Suspended detritus, larval forms, or plankton are trapped in ciliary currents, in mucous sheets, or by mechanical screening and are then transferred to the mouth.

4. Homing on Food

Although many gastropod species, particularly carnivorous prosobranchs, can home on food materials in still water over distances of a meter or more, the data for herbivores are less clear. Study of locomotor behavior of the herbivore *Stagnicola reflexa* (Bovbjerg, 1965) has shown (1) that in a Y-maze with homogenized food substance (alga *Spirogyra*) restricted to one arm of the maze animals chose both arms approximately equally often, (2) that only if the food substance extended into the leg of the Y did more snails choose the correct arm, and finally (3) that snails placed at the center of a straight runway in still water, with food at one end only, chose both directions for travel equally often. Distance chemoreception may therefore not be important in food location for this herbivore (but see Bovbjerg, 1968; localization of food by *Lymnaea*).

Although Frings and Frings (1965) found that *Aplysia juliana,* buried in the sand, would emerge immediately when *Ulva* was placed in its aquarium, it may be that currents and small food particles contributed to elicitation of the response. This is particularly likely because the aquaria used were quite small (1 gal capacity). Preston and Lee (1972) have found that while

Aplysia californica can locate *Ulva* in still water, this capability is not highly developed. The distances involved were "at most, only one or two body lengths," and the times taken to initial food contact were of the order of 30 min. Uncontrolled visual cues may have been contributing factors in these experiments. On the other hand, Preston and Lee found that subjects aroused by food chemicals before placement in a water current turned directly upstream. They then continued toward the source despite the absence of food substance from the water stream. On the basis of these and other experiments. Preston and Lee concluded that sensory structures in the region of the mouth and anterior tentacles (but not the rhinophores or osphradium) detect food chemicals and establish an aroused condition in which the animal turns into prevailing currents. It is as yet unclear whether homing on algae in *Aplysia* is brought about by continuous reference to chemical gradients or whether the presence of food in the water elicits turning into the current and forward locomotion. Experiments still need to be performed to separate the rheotaxic and chemotaxic components of food-seeking behaviors.

As has been pointed out by Owen (1966), it is difficult to distinguish between taste (contact chemoreception) and smell (distance chemoreception) in aquatic organisms. In this regard, however, it has been noted that *Fusitriton* (C. Eaton, personal communication) and *Melongena* (Turner, 1959) "gang up" on an individual prey, probably employing distance chemoreception. When one begins to feed on prey organism, numerous others from the nearby environment sense the feeding activity, most likely by "smell," and immediately move to join the attacker. Furthermore, *Melongena* will do so in preference to attacking other killed organisms made available to it nearby. The evidence available suggests that the attackers secrete something into the surrounding water which acts as a very potent attracting agent for others of the same species. Although none of the attracting agents either from the attackers or from the prey are known, substances such as oxaloacetic acid, trimethylamine, tetramethylammonium hydroxide, and a number of amino acids are effective at extremely low concentrations; e.g., 0.0001 % trimethylamine attracts *Bullia* (Brown, 1961).

Snyder and Snyder (1971) found that *Fasciolaria tulipa,* a predacious cannibalistic marine prosobranch, may have a pheromone-mediated feeding behavior. When a large individual crosses the downstream odor trail of a smaller one, it turns toward the smaller animal and attempts to capture it. Contrarily, if the sizes are reversed, the downstream individual attempts to escape rather than capture. Thus conspecific size discrimination seems to rely on odor stimuli. The basis for discrimination is probably quantitative rather than qualitative, since six small snails placed together upstream produced fewer capture responses and more escape responses in a medium-sized respondent than did a single small snail.

Some gastropods preferentially follow mucous trails laid down by other members of their species or in other cases follow prey species. A most striking example of the information that can be gleaned by a snail from a mucous trail has been discovered by Hall (personal communication). When *Littorina* encounters a trail made by another member of its species, it turns to follow the trail. Surprisingly, however, it does not turn in both directions along the trail with the same likelihood. Instead, it turns to follow the trail-making snail in a statistically significant percentage of trails. This occurs quite regardless of the sex of either follower or trailmaker, and independent of the angle of approach to the trail or the left–right direction required for following. Although the sensory mechanism underlying this process is unknown, observation of the perpendicular orientation and tentacular movements of the subject when the trail is first encountered suggests that an odor gradient along the 1-cm distance sampled by two tentacles may be used. Such a mechanism, if present in this and other species, may be important in homing (see below).

V. LEARNING STUDIES

Studies of learning in gastropods have evolved through three stages over the past 80 years. The first 50 years were occupied almost entirely by purely behavioral experiments. Some of the accounts of learned behaviors out of this period were dramatic (e.g., see the Appendix for a reported instance of a snail that learned to recognize its master's voice!) but insufficiently analytical to instill much confidence. Many others were carefully designed experiments and reported in enough detail to be relevant and useful in the present context. The second period corresponded with the discovery and then exploitation of the special potential for neurophysiological experiments of the isolated nervous systems of gastropods with their unusually large nerve cell bodies. The pioneering studies of Mme. Arvanitaki and subsequently many others showed that the neurons in the isolated ganglia of *Aplysia* and then of other gastropods were extraordinarily useful objects of electrophysiological study. As knowledge of the input and output connections to many gastropod neurons has become available, along with information about the normal or spontaneous electrical activity in the cells, interest has focused on the possibility of producing changes in neuronal electrical activity that are supposed to parallel behavioral changes which accompany learning. Recently, it has become clear that there is no longer the need to limit studies to either purely behavioral or purely electrophysiological aspects of learning but instead is possible to combine the two. While both behavioral and electrophysiological studies of gastropod learning continue, the combination of the two is now attracting most attention and in the long run seems most likely to illuminate the fundamental mechanisms of learning.

A. Purely Behavioral Studies of Learning

Several aspects of the behavior of gastropods have been modified either by habituation or by conditioning schemes. Often, the sensory stimulation has been tactile or photic and the motor responses have been withdrawal of either the gills or some part of the gills. Other frequently studied motor responses include withdrawal of the tentacles, expulsion of air from the lung, withdrawal of the whole animal into its shell, or movements of the mouth associated with feeding (Table II). The majority of experiments to date have involved habituation of responses, but a few have been directed at operant and classical conditioning and an even smaller number at discrimination learning tasks.

1. Pieron's Studies (1905–1915)

Using the pond snail *Lymnaea stagnalis* and the common shore winkle *Littorina,* Pieron gave the first compelling evidence that gastropods were capable of habituating to repeated stimulation. In every case, his work took advantage of the reflex withdrawal elicited by shadows cast over them or mechanical vibrations transmitted to them. Both species responded by partial or complete withdrawal of tentacles and head into the shell. He noticed that if the stimulus was repeated several times the strength of the withdrawal responses declined. After a variable number of stimulations, usually ten to 15, responses no longer occurred.

Virtually all of Pieron's experiments follow the same basic pattern. A snail was placed in a partially closed glass container in pond water from its home environment. The incident light level was maintained at a constant, relatively high level, and shadow stimuli were delivered by briefly interrupting the light path from the source with an opaque body. The duration of the stimulus was quite brief, of the order of 1 sec. The response of the animal, as mentioned above, was withdrawal of the forward part of the body toward or into the shell. It is unclear whether the light sensitivity revealed by these phenomena involves the eyes or other undescribed photoreceptors. In view of Medioni's (1959) finding that cutting off the eyes of *Lymnaea* does not eliminate reflex withdrawal, it is at least clear that eyes are not required.

In his earliest reported studies (1910), Pieron noted a fact which may be of wider significance for experiments using gastropods. Individuals from snail populations taken from different natural environments showed different sensitivities to shadow stimulation. For instance, snails collected from ponds which were richly lined with weeds and grass but not shaded by trees

Table II. Gastropod Learning—Behavioral Studies

Author(s)	Animal	Behavior	Result
Pieron (1910, 1911, 1913)	*Physa* and *Lymnaea*	Reflex withdrawal into shell when shaded	Habituates (not controlled for sensory adaptation)
Dawson (1911)	*Physa*	Reflex withdrawal into shell when disturbed	"Tamed" (habituated?) snails less sensitive to disturbances
Crozier and Arey (1919)	*Chromodoris*	a. Reflex withdrawal of rhinophores to tactile stimulus	Habituates (not controlled for sensory adaptation)
		b. Reflex withdrawal of gills to tactile stimulus	Initially habituates (no controls)
		c. Reflex gill withdrawal to shading	Habituates (not controlled for sensory adaptation)
Buytendijk (1921)	*Lymnaea*	a. Righting response after being turned upside down	(1) Latency to righting declines (i.e., habituation occurs) provided animal is handled carefully; (2) latency remains long if animal is handled roughly (may be conditioned by aversive stimulation)
		b. Crawling backward out of narrow tube	No improvement in performance
Humphrey (1930)	*Helix*	Reflex withdrawal into shell to mechanical stimulus	Habituates
Bruner and Tauc (1965b)	*Aplysia*	Reflex tentacle withdrawal to water-drop stimuli	Habituates
Cook (1971)	*Lymnaea*	Withdrawal by forward and downward movement of shell	Habituation and dishabituation; visual and mechanical stimuli produce independent effects
Peretz (1970)	*Aplysia*	Reflex gill withdrawal to water-drop stimuli (isolated gill preparation)	Habituates
Pinsker et al. (1970)	*Aplysia*	Reflex gill withdrawal to tactile (water-jet) stimuli	Habituates
Abraham and Willows (1971)	*Tritonia*	Escape-swimming response to salt stimulation	Duration of response declines (habituates?)
Carew et al. (1972)	*Aplysia*	Contraction of siphon, gill, and mantle shelf to tactile stimulation of siphon	Habituates (long-term)

Table II (*Continued*)

Author(s)	Animal	Behavior	Result
Thompson (1917)	*Physa*	a. Feeding response to simultaneously paired tactile stimulation and food presentation	Classical conditioning after 250 trials (incomplete sensitization controls)
		b. Negotiates U-shaped labyrinth to surface of aquarium	Little or no improvement in performance over trials
Sokolov (1959)	*Physa*	Escapes from one arm of H-shaped apparatus to other in response to onset of light paired with noxious chemical stimulation	Classical conditioning (incomplete data available)
Jahan-Parvar (1970)	*Aplysia*	Mouth-opening response to paired light flash and food presentation	Classical conditioning (incomplete controls)
Abraham and Willows (1971)	*Tritonia*	Escape-swimming response to paired salt and tap-water stimuli	Sensitization occurs; no evidence for classical conditioning
Wells and Wells (1971)	*Physa*	a. Withdrawal response to paired "light-off" or mechanical shock with electric shock	Sensitization occurs; no evidence for classical conditioning
		b. Movement across light–dark, smooth–rough patch in dish leads to shock	No improvement in performance
Lickey and Berry (1966) and Lickey (1968)	*Aplysia*	Feeding response	Improved discrimination between food and nonfood objects
Lee (1969, 1970)	*Aplysia*	a. Movement to particular location of aquarium to obtain raised water level in aquarium	Animal enters required area more often than yoked control
		b. Reaches head up to successively higher levels in slot to obtain raised water level in aquarium	Animal reaches higher and more often than yoked control
Garth and Mitchell (1926)	*Rumina*	T-maze reinforced by darkened goal box	Time in maze declines; number of errors declines
Fischel (1931)	*Lymnaea* and *Ampullaria*	a. Maze requiring right-hand turn; "correct" performance coaxed by stimulation with fine brush	Little or no improvement

Table II (*Continued*)

Author(s)	Animal	Behavior	Result
		b. Passage across jagged tin barrier leads to poking with pin	With training, contact with tin elicits reflex withdrawal (no sensitization controls)
		c. Three-way choice at end of runway, one leading to shock	No "learned" improvement in performance
Cook (1969), Arey and Crozier (1921), Hewatt (1940), and others	*Various limpets and snails*	Homing to old "scar" on rock when disturbed or displaced or when rock is rotated or local environment restructured	Homing performance generally unaffected except when animal is placed in "unfamiliar" or distant region; memory of local area or homing on odor cues
Verlaine (1936) Fischer-Piette (1935)	*Natica Nucella*	Boring holes in prey	Improved boring location chosen with experience

around the shore had a greater sensitivity to shadows than did snails from stream beds which were practically devoid of weeds and grasses but were overshadowed by numerous trees along the banks. His interpretation of this difference was that the snails from the shaded streams experienced regular shadowings owing to wind blowing through the trees along the riverbanks. This, he supposed, produced a very long-term habituation to shadows.

He then attempted to produce a similar kind of long-term effect in the laboratory by subjecting animals to daily series of shadings for as long as 3 weeks. He found, however, that habituation, i.e., a reduction in the number of shadings required before no withdrawal response was produced, could be detected only over the first few days at best and in one instance not at all. After that, the animal's sensitivity to shadows no longer declined and in fact remained at a fairly high level indefinitely, permitting the conclusion that during the nearly 24-hr period between series of shadows the memory of those events declined almost to the point of vanishing so that the animal could not profit from the previous day's experience. Apart from the explicit conclusion about the duration of the "memory" of the previous day's trials, the more general conclusion is suggested that features of the laboratory environment may prevent forms of learning which develop easily in the natural environment.

In 1910, Pieron also found that the response level to shadows in *Lymnaea* declined more rapidly over a series if the animal had been subjected to similar stimulus series at a time in the recent past, i.e., within approximately 1 hr. Each animal was subjected to a series of 15 shadows separated from one another by 1 sec, each being $\frac{1}{4}$ sec in duration. The number of successive withdrawal responses before habituation was recorded. After a variable interval ranging from 20 sec to 2 hr, the same sequence was repeated and the number of responses noted. An improvement in performance on the second sequence was measured by dividing the number of responses in the first series into the difference between the number of the first and the second series. After a 20-sec interval, it was found that "savings" computed this way amounted to just over 90% and that over the period of an hour the savings declined to 18%. The decline was not linear, and most of the reduction occurred in the first few minutes (Fig. 9).

In 1911, Pieron reported that the rate of habituation of the withdrawal response was a marked function of the interstimulus interval and that the optimal interval for rapid habituation was quite different in *Lymnaea* and *Littorina*. He found, for instance, that habituation in *Lymnaea* was very poor with intervals of less than a second or longer than a minute and that the optimal interstimulus interval was about 4–8 sec, whereas in *Littorina* the optimal interval was approximately 1 min. Thus two animals in the same class had optimal interstimulus intervals for habituation to the same stimulus that differed by an order of magnitude. Furthermore, he pointed out that the rate of decline of the fully habituated state was more rapid in species that habituated at shorter interstimulus intervals. Specifically, *Littorina* reacquired the response after 5 min as well as *Lymnaea* did after 40 sec.

In a comprehensive comparative report in 1913, Pieron restated these results and in addition outlined his reasoning for supposing that these phenomena were in fact central nervous in origin and not simply the result of sensory adaptation or motor fatigue. His principal argument against motor fatigue was that a fully habituated animal responded by withdrawal to a single stimulus applied via another sensory route, *viz.,* a mechanical shock, with as much vigor as its initial unhabituated response to the shadow. The argument would be that precisely the same muscles drive the withdrawal movements regardless of the source of the stimulation. Therefore, since the second pathway elicits the response in an animal that has been fully habituated to the first, muscle fatigue cannot be responsible for the observed waning of the withdrawal reflex. A different argument was used to discount sensory adaptation as the cause of the observed habituation. Pieron pointed out that the stimulus in all of the shading experiments did not consist of excitation in the usual sense of the word but rather in a brief respite from on-going stimulation namely, the high background illumination level. As he pointed out, it is difficult to imagine how a brief reduction in the incident light level could "fatigue" a receptor, especially since the stimuli were all very short, ranging from $\frac{1}{4}$ sec to 1 sec in duration. Central to the validity of this argument is the unwarranted assumption that the receptors involved are not excited by reductions in incident light intensity.

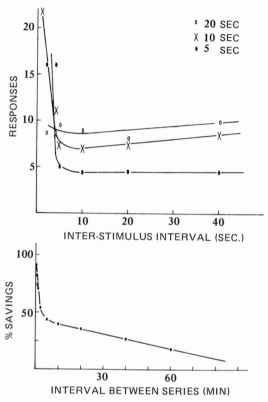

Fig. 9. Habituation of withdrawal response to shading in *Lymnaea* (Pieron, 1910, 1911). Upper graph: The effect of interstimulus interval or number of successive withdrawal responses before shading failed to produce withdrawal. Whether tested 5, 10, or 20 sec after the first failure to withdraw, the optimal interstimulus interval for habituation of withdrawal was 4–8 sec. Lower graph: With two equal series of shadings separated by a variable interval, the number of responses to the second series was usually less, indicating some measure of retention. Savings were determined as the number of responses to the first series divided into the difference between the number of responses in the first and the second series. Retention fell rapidly over 10 sec after the first series and dropped below 20% within 1 hr.

2. Additional Studies of Habituation

In a brief commentary on "some psychic phenomena of *Physa*," Dawson (1911) found that the sensitivity of this pond snail to localized mechanical stimulation depends very much on the recent history of the animal. For instance, a "tamed" snail (from an aquarium in a busy laboratory) that had been regularly handled and subjected to mechanical stimulation of one sort or another would normally not withdraw into its shell or even withdraw its tentacles when approached or picked up, while an animal that had been kept in a different aquarium in a room where it was essentially undisturbed for long periods would withdraw completely in response to the same stimuli. Furthermore, she noted that by reversing the environments of these two animals for an unspecified length of time their relative sensitivities to mechanical stimuli were also reversed.

In the course of describing the sensory reactions of the Bermudan nudibranch *Chromodoris zebra*, Crozier and Arey (1919) noted decrements in the withdrawal response of the rhinophores and the anal gill tuft. *Chromodoris* has a pair of rhinophores at the anterior end of its dorsum and a circular tuft of gills surrounding its posteriorly placed anus. These authors did not study the reactions of either of these organs systematically but commented anecdotally on the pattern of responses to repeated stimulation.

They noted, for instance, that 50 successive mechanical stimuli applied directly to one side of one of the rhinophores with 10-sec interstimulus intervals produced a decrease in the amplitude in contraction, but they also recorded that the rhinophore was still capable of reacting. In regard to the gill, mechanical stimulation of the brim of the anus with a glass rod initially produced a complete withdrawal of all the gills. The time required for subsequent reexpansion of the gill at first decreased (Fig. 10) over six or eight stimuli but then increased until it approached values similar to those encountered initially. Unfortunately, these latter studies are described very

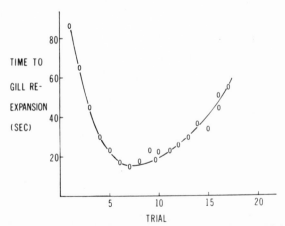

Fig. 10. Gill withdrawal and reexpansion time in *Chromodoris* over trials (Crozier and Arey, 1919). Although the time to reexpansion of the gill declined rapidly over five to ten trials, indicating a possible habituation the duration of the response, the time to reexpansion increased again over the next ten trials, indicating that sensitization may occur with a delay.

sketchily, and none of the details of interstimulus interval or the precise nature of the stimulus are provided.

They observed also that the gill plume of *Chromodoris* was immediately withdrawn when a shadow passed over the gill. Unlike the rhinophore response, however, the withdrawal reflex declined rapidly with gill stimuli at 30-sec intervals. Indeed, as few as two or four such stimuli were sometimes sufficient to produce complete habituation. Typically, the habituation of gill withdrawal was highly variable. For instance, one individual responded 20 times in succession to shadows with no apparent reduction in response strength.

According to these same authors, another nudibranch, *Facelina goslingi,* is also reactive to shading, although the shade must extend over the head to produce a response. The withdrawal takes from 0.6 to 1.0 sec and apparently has a 30–60-sec refractory period during which no shading response can be elicited. This reflex was especially difficult to fatigue, a fact which may have been related to the long refractory period.

If the snail *Lymnaea* is lifted from the substrate and then replaced upsidedown, it reemerges from its shell after a variable period of time and then rights itself by reaching to the substrate, attaching its foot, and pulling over its shell. Buytendijk (1921) demonstrated that habituation of the withdrawal response is interfered with by rough post-trial handling of the subject. He measured the time taken on successive trials to return to an upright, foot-attached position and found that if, after righting, the subject was roughly pulled from the glass and tossed back into its aquarium to await the next trial, no reduction in time withdrawn occurred. If, on the other hand, following a successful recovery, the animal was carefully guided to the edge of the glass and slid gently onto the plants in its home tank, a marked reduction in response latency occurred with successive trials (Fig. 11). The author concluded that the snail formed an association between emergence and rough handling, and consequently it remained withdrawn for longer periods. At the same time, however, it was habituating to being placed upside down. The net result was a varut, iable bon average, relatively long response latency. This very simple and natural response invites further study, particularly since both habituation and conditioning may be involved.

This same author reported also an unsuccessful attempt to train *Lymnaea* to crawl backward from a blind tube. Each subject was placed head first into the open end of a close-fitting glass tube. Locomotion other than forward is uncommon and perhaps difficult for *Lymnaea,* and it always proceeded down the tube to the blind end. Upon reaching it, turnaround was impossible and the only way out was by crawling backward. Although a rather stereotyped pattern of movements often developed (Fig, 11), no reduction in time or path length in the tube was apparent. The length of the tube used was not given. The simplicity and elegance of this experiment suggest that further

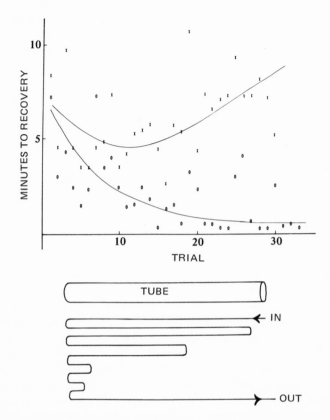

Fig. 11. Upper: Recovery time from placement in upside-down position in *Lymnaea* (Buytendijk, 1921). Recovery time first declined then increased in experiment when snail was roughly handled after recovery (*X*). Recovery time declined to a few seconds and remained short when recovery was followed by gentle transferal to home aquarium in later experiment (*O*). Lower: Path followed by *Ampullaria* after placement in narrow, blind glass tube from which exit could be gained by crawling backward. No systematic decline either in path length or in time to escape was observed, and the typical sequence of movements resembled the example shown. Collision with end of tube elicited successively shorter backward crawling sessions until snail was finally stopped at end. At this point, subject usually crawled backward all the way out.

attempts should be made, using shorter tube lengths and also perhaps other gastropod species for which rearward pedal locomotion may be more normal.

In a series of experiments taking advantage of the same type of withdrawal response described by Pieron, but using the land pulmonate *Helix albolabris,* Humphrey (1930) demonstrated that the response not only habituated but also recovered with rest, dishabituated, and had several other properties which are now associated with habituation in other animals. The experiment consisted of placing snails individually upon a wooden platform, which could be given a brief, measurable longitudinal jerk via the electromagnet of a bell-ringing device, and then subjecting each to a series of jerks separated by 2-sec intervals. He noted the extent to which the tentacles were withdrawn and whether the animal withdrew into its shell. Ten snails were used. Considerable variability could be seen from individual to individual in terms of the rate of habituation, but in virtually every case responses failed after a surprisingly few stimuli (Fig. 12). The animal could be made to respond again by interposing an extraneous stimulus (a heavy blow on the platform) between two normal stimuli.

Dishabituation itself was found to habituate. One snail, for instance (No. 7, Fig. 12), habituated fully to the jerks, dishabituated several times to the dropping of the ball, but then became less and less sensitive to the dishabituatory stimulus until finally it produced only very slight responses and the amount of recovery from habituation to the jerk stimulus was no longer perceptible.

An additional finding from these same experiments was that the habituated state deepened with continued stimulation after failure of the response. For instance, the amount of recovery after a rest depended in part on the number of habituatory stimuli that preceded the rest. Evidence on this point can be seen in the results of experiments with snail No. 20. After 18 stimuli, only the first of which produced a response, a rest was given which was followed by a response and a total of 102 additional stimuli. After the next rest, no obvious response occurred, and 118 additional stimuli produced no response. The suggestion is that the habituated condition was deepened by the 102 stimuli of the second sequence to the point that the rest was insufficient to permit a reappearance of the response after the second rest.

Finally, in one snail, Humphrey noted that if he reduced the interstimulus interval from 2 sec to 1 sec the withdrawal response no longer habituated and instead remained strong over 182 stimulations. This observation correlates with earlier results of Pieron and of Crozier and Arey (mentioned above) where the optimal interstimulus interval for demonstration of habituation was not necessarily the shortest interval.

This study with *Helix* represents the first instance in the gastropods in which dishabituation was demonstrated, and it strongly encourages the view that the response decrement was due to a change in central neurons rather than a decline in the efficacy of the sensory or motor components.

Pieron's (1911) early studies describing habituation in *Lymnaea* were confirmed and greatly extended by Cook (1971), in a systematic analysis of withdrawal in response to mechanical and photic stimulation. Both frequency of responses and their amplitudes were shown to decline with repetition of stimuli. However, latency to withdrawal did not change appreciably. Unlike

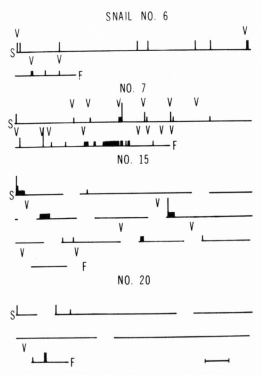

Fig. 12. Habituation and dishabituation of withdrawal in *Helix* in response to mechanical shocks (Humphrey, 1930). Platform upon which snail crawled was jerked every 2 sec over entire period of experiment. Start and finish of each experiment are indicated by *S* and *F*, respectively. A complete withdrawal is indicated by the tallest vertical lines, and lesser responses are indicated by shorter marks. No vertical mark rising from the time axis indicates that no response occurred. Dishabituatory stimuli (steel ball dropped onto platform) were applied at *V* marks. Snail No. 6 stopped responding after only two complete withdrawals at beginning of experiment and only responded occasionally thereafter until dishabituatory stimuli were applied. There was a slight and short-lived recovery in response to jerks after each. Snail No. 7 habituated after one response. It then recovered in most cases for one or a few trials after dropping of the steel ball until the last four such dishabituatory stimuli. Then recovery was no longer apparent. Snail No. 15 habituated after several trials. Periods of rest (blank sections on time axis) were usually followed by a partial recovery in responsiveness. Such recovery was blocked in snail No. 20 when the number of trials was increased. Responses after the first rest period were followed by 102 stimuli. After the next 30-sec rest period, no recovery occurred. Likewise, no recovery was evident after the next long stimulus sequence and rest. A dishabituatory stimulus during the last rest period led to a few responses. Time calibration: 40 sec.

the results with vertebrates, habituation in *Lymnaea* proved to be insensitive to the temporal patterning of shocks or to changes in stimulus duration.

A significant aspect of this study was a comparison of habituation to different classes of stimuli. In particular, mechanical (mechanical shocks and vibration) and photic stimuli (light-off and moving shadows) were compared. Rate of habituation to mechanical shocks or vibration was more rapid than when light-off or moving shadows were used. Habituation to mechanical shocks or vibration did not influence subsequent responses to either light-off or moving shadows and *vice versa*. Stimuli affecting different sensory modes apparently produced habituation independently.

Habituation proved not to be stimulus specific. Repetition of mechanical shocks simultaneously produced habituation to vibrations and *vice versa*. As well, light-off stimuli habituated the animals to moving shadows. Cook's (1971) data on the time course of recovery from habituation to light-off stimuli suggest that recovery was nearly complete after about 5 min. This compares favorably with the rapid decay of habituation noted over the first 2–5 min in Pieron's data (Fig. 9) for a similar response in the same animal.

3. The Problem of Habituation and the Peripheral Nerve Net

A question arises regarding the neuroanatomical site for habituation phenomena. Interpretation is made much more difficult by the apparent fact that gastropod mollusks have, in addition to a central nervous system, a peripheral nerve not diffusely scattered throughout most peripheral tissues which is capable of mediating quite widespread sensory–motor responses. It is now certain that some of the behavior decrements found in gastropods depend on neural events which occur independently of the central nervous system. While this does not reduce the intrinsic interest of the phenomena, it could make the experimental problems of determining the underlying mechanisms more difficult. Little work has been done on the peripheral nervous system of gastropods, but there is evidence from both histological and physiological points of view for its existence.

Histological evidence suggests either a conducting glial–interstitial cell network (Nicaise, 1967; Nicaise *et al.*, 1968) or a peripheral nerve net (Gorman and Mirolli, 1969; Retzius, 1892; Veratti, 1900; Born, 1910; Bethe, 1903). The electron microscope shows a continuous cell network surrounding the neurons of the central nervous system and extending out to the periphery even to the extent of interconnecting muscles. Provided that these elements carry electrical activity (contacts resembling synapses are seen from this network onto neuron and muscle cells), they may play the role of a parallel conducting system passing excitation from central neurons and peripheral sensory sources to other central neurons, and possibly even to muscles.

Light microscopic studies imply only that extensively branched neural netlike processes pass throughout the muscle fibers of the foot, body wall, and gut. Their functional roles as neurons cannot yet be stated with full confidence.

A second line of evidence pointing to the existence of a peripheral nerve net is behavioral. *Isolated* gill crowns of the Bermudan nudibranch *Chromodoris zebra* consisting of 12 distinct plumes rising from a ridge surrounding the anal aperture retain their reflex withdrawal capability in response to tactile stimulation and, in a reduced form, even to shading (Arey and Crozier, 1919; Crozier and Arey, 1919). Furthermore, it is a common observation of workers with gastropods that isolated pieces of tissue are capable of extensive spontaneous and reflex movement (Roach, 1967), and animals from which the central ganglia have been removed show locomotor activity and even contraction of extensive areas of body wall musculature in response to local stimulation.

Experiments with various muscles in diverse species in which brain cells have been stimulated and responses recorded photo- or mechanoelectrically yield contradictory results. On the one hand, Hoyle and Willows (1973) found that the peripheral responses to brain cell stimulation in *Tritonia* were delayed as much as several hundred milliseconds (beyond that accounted for by expected conduction delays). Furthermore, responses to a series of identical central stimuli were extremely variable in magnitude and extent. Finally, peripheral summation and inhibition were observed. All these findings are most readily compatible with the existence of a peripheral nerve network that mediates transmission from brain neurons to muscle.

On the other hand, similar experiments (see also below) with motor neurons in the parietovisceral ganglion of *Aplysia* (Kupfermann *et al.,* 1971) provide evidence that these centrally located neurons directly innervate muscles that cause withdrawal of the gill as a whole. Excitatory, depolarizing potentials recorded in what are presumably muscle fibers are phase locked and on a one-for-one basis with spikes in the neuron. Yet this same study and an earlier one of Peretz (1970) both clearly implicate a semiautonomous peripheral nerve network as the mediator of the local reflex responses of the individual pinnules which make up the gill.

Further evidence for direct innervation of muscles by neurons whose somas are located in the ganglia comes from the findings of Kater *et al.* (1971). When recorded extracellularly, the muscles responsible for positioning of the odontophore and buccal mass and for withdrawal into the shell in the basommatophoran snails *Helisoma* and *Planorbarius* exhibit phase-locked, one-for one, excitatory potentials when appropriate neurons in the buccal ganglion fire impulses. Evidence from similar discrete muscles in the opisthobranch *Tritonia* is also compatible with this interpretation (Fig. 13).

Taken together, these apparently contradictory results suggest that

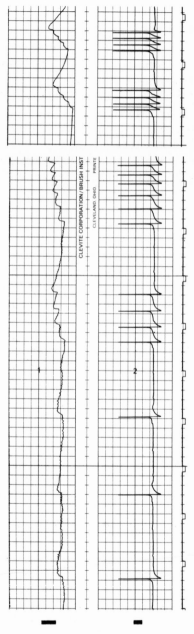

Fig. 13. Simultaneous intracellular recordings from buccal ganglion neuron (lower trace) and muscle fiber of buccal mass in *Tritonia*. Each action potential in the neuron (stimulated by passing depolarizing currents through recording electrode) was followed after approximately 30 msec by a 3–5-mv depolarizing potential in the muscle. Muscle potentials summate, but spikes were not seen, although shortening could be observed to take place in muscle in response to short spike trains. Muscle fiber resting potential was 30 mv. Voltage calibrations were 10 mv. Time marks indicate seconds.

discrete anatomical muscles made up of parallel fibers with well-defined attachment sites (e.g., muscles of the buccal mass) may be directly innervated, while diffuse muscles capable of contracting in several directions (e.g., body wall or foot) are under the direct control of the peripheral nerve network and only indirectly driven by centrally located neurons. Intermediate types such as those found in gills, which, although not morphologically distinct, are nevertheless arranged sufficiently in parallel to permit contraction in essentially one direction only, may be innervated directly, indirectly, or via both routes in parallel.

   a. Habituation in the Absence of a Central Nervous System. The importance of defining the role of a peripheral nerve net is emphasized by two nearly simultaneous reports of habituation in *Aplysia*. Pinsker *et al.* (1970), Kupfermann *et al.* (1970), and Castellucci *et al.* (1970) reported that the gill withdrawal reflex of *Aplysia* relied directly on sensory and motor neurons in the parietovisceral ganglion. These experiments are discussed in greater detail below. Another study by Peretz (1970) showed that a similar habituation occurred in the gill with local stimulation, in the complete absence of the parietovisceral ganglion. See also Lukowiak and Jacklef (1973).

   In Peretz's experiments, the gill and a small patch of nearby skin were removed and withdrawal movements of single pinnules (the gill is made up of 12–16 such subunits) were recorded with a mechanoelectric transducer. Seawater dropped onto the pinnules from a 14-cm height evoked withdrawal movements that became less with successive stimuli. The withdrawal responses declined in amplitude more rapidly when short interstimulus intervals were used and more slowly with longer intervals over the range of one drop per 0.5 min through one drop per 2.5 min, with a maximal decline to approximately 50% of the control level over a time of the order of 15–45 min (Fig. 14). The response recovered either after a period of rest or following direct electrical stimulation of the pinnule or the ctenidial nerve. Habituation produced this way endures for as long as 2 hr. Histological study of the gill revealed several types of neurons, some located in clusters in the bases of the pinnules. Others with axonlike processes were found along the epithelial margins of the pinnules. These may have been sensory neurons. Still others were located in the gill musculature. Peretz noted that impulse activity could be recorded both from the surface of the pinnule and from the cut end of the ctenidial nerve, and he suggested that the recorded activity was produced by elements of the local nerve network.

   A similar but not identical withdrawal response of the whole gill can be elicited by sensory stimulation of nearby regions of the mantle shelf and siphon. As has been shown by both Kupfermann and Kandel (1969) and Peretz (1969), this latter response is probably driven by neurons in the parietovisceral ganglion. Recent evidence (Kupfermann *et al.*, 1971) confirms that

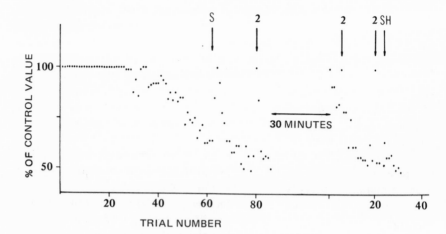

Fig. 14. Habituation of gill pinnule reflex in the absence of the central nervous system in *Aplysia* (Peretz, 1970). Water drops applied at 5/sec to the gill elicited a withdrawal which declined to 50% of control value after 80 stimuli. Spontaneous gill movement (*S*), pairs of water drops (*2*), and electrical shock to the pinnule (*SH*) all produced marked, temporary recovery in response. Likewise, a 30-min rest was followed by recovery and then rehabituation at a more rapid rate.

there are two parallel and independent pathways by which gill movements can be produced, one which is peripherally located and consists most likely of sensory and motor neurons in the gill (although direct excitation of muscle by the mechanical stimulation has not been ruled out) and the other involving both sensory and motor neurons of the abdominal ganglion. These authors drew two conclusions. First, the neural pathways through the abdominal gangilon are necessary and sufficient for the gill withdrawal reflex, and in particular the motor neuron L7 innervates the gill muscles directly without engaging the peripheral nerve net. Second, the nerve net probably controls the reflex pinnule response, as distinct from the gill withdrawal reflex. But see Lukowiak and Jacklet (1973).

b. Habituation of Escape Swimming. A fundamentally different kind of behavior, one which involves the central nervous system in a more complex way, was studied by Abraham and Willows (1971) and Willows (unpublished observations). The behavior is an escape-swimming sequence in the nudibranch mollusk *Tritonia.* The swimming response, which is normally elicited by contact with starfish, or in the laboratory by salt crystals or soap, consists of a sequence of from three to ten swimming cycles lasting a total of approximately 20–70 sec. With repetition of stimulation at intervals of 5–25 min, the following was found: (1) The likelihood of initiation of swimming responses and the durations of response latencies did not increase or decrease appreciably, provided that the stimulus was of moderate or strong intensity. Under these circumstances, swimming virtually always occurred and did so with a latency of 5–10 sec. (2) If the stimulus strength was minimal (a few crystals of salt), then the likelihood of initiation of swimming was reduced during the first one or two trials and only rose to a maximum value after two or three presentations (Fig. 15). (3) The duration of each response, measured as the number of swimming cycles, declined to approximately 50 % of control values over approximately ten trials. (4) The rate of habituation of the response increased as the interstimulus interval decreased. With moderate and strong stimuli, it was apparent that the neural apparatus responsible for initiating the escape-swimming sequence does not change in sensitivity to the appropriate stimulus over many trials (200 were tried). It seems instead that other components which control the duration of the response may undergo a change. The neural events underlying the habituation are almost certainly central, since the neural activity appropriate to swimming can be produced in the isolated central nervous system, and under these circumstances the response duration is normal (Dorsett *et al.,* 1969, 1973). Furthermore, dishabituation occurs when a novel stimulus (such as a mechanical disturbance) is presented between trials.

4. Learning Other Than Habituation

Behavioral studies of more complex associative forms of learning in gastropods have included both successful and unsuccessful classical conditioning experiments and studies of maze performance, of operant response differentiation, and of passive or inhibitory avoidance behavior (Table II).

a. Classical Conditioning. The oldest and most widely quoted study recording classical conditioning was that of Thompson (1917) with the freshwater snail *Physa*. *Physa* cruises in the surface tension in search of food. It accomplishes this by hanging upside down, its foot placed directly against the water–air interface, with the center of its foot depressed into a boatlike concavity. Apparently, the force of the surface tension plus the buoyancy conferred by the displaced water is enough to carry the weight of the snail. Thompson noticed that when the mouth region was touched with a bit of lettuce or with its own mucus or even a glass rod, the snail usually made biting movements. Her conditioning scheme was to pair simultaneously the application of mechanical pressure (a brief pulse of pressure to a region of the foot remote from the mouth) with the delivery of a small bit of lettuce to the mouth region. Initially, she showed that application of pressure to the foot alone rarely if ever caused the feeding response. On the other hand, of the six snails trained in a series of 250 paired trials, then given a 48-hr rest period, all showed several feeding responses when stimulated with pressure alone. In particular, one snail responded vigorously to pressure alone in each of the nine test trials. Since the last two of these nine were separated from the first seven by an interval of 48 hr, she concluded that the effect of the training persisted for 96 hr. Similar results were obtained from the other five snails.

This study can be criticized on the grounds that there are no adequate controls for sensitization. Although the author noted that 20 applications of pressure alone produced only four responses in six naive snails, the conclusion is not warranted that pressure alone has no sensitizing effect after 250 paired trials. No indication is given of the interstimulus intervals either for the pressure stimuli alone or for the combined pressure and food stimuli, and so it cannot be determined whether the total stimulation over time was the same in both cases. Probably more importantly, the number of responses to the combined pressure and food stimulation in the conditioning training was much reduced during the first 100 trials by comparison with food-alone controls. This implies that the pressure by itself had an inhibitory effect on the feeding response initially, an effect which declined over the first 100 trials. It is therefore not surprising that 20 pressure-alone trials produced few responses. It may be that after about 100 trials the inhibition due to pressure declines and thereafter sensitization occurs. This aspect of the experiment clearly needs further study.

Sokolov (1959) tried to condition *Physa* to crawl from one water-filled compartment into another when a light went on, by pairing the onset of the light with the delivery of an irritating chemical stimulus (0.2% KCl). The H-shaped apparatus consisted of two identical chambers joined by a connecting arm. The two upright tubes were isolated from one another by a light barrier. Following onset of the light on the side of the barrier where the snail was located, KCl flowed into the same chamber. The animal could only escape from this unpleasant situation by crossing to the other chamber. Sokolov's conclusion that the conditioned reflex is established with a speed that does not differ from that of other animals (its first display appearing in less than eight combinations of the CS and US with consolidation in less than 30 combinations) seems to be somewhat overstated. The data provided refer to only two out of the nine snails tested in detail, and even here the evidence for conditioning is not impressive.

A preliminary study by Jahan-Parvar (1970) using the mouth-opening response (MOR) of *Aplysia* is also equivocal. A light was flashed $\frac{1}{2}$–1 sec before presentation of a small piece of seaweed into the mouth area, and responses were scored in terms of the subject's food-searching and -finding movements. The paired stimuli were delivered ten times with 5-min intertrial intervals, and then the subjects' responses to light alone were tested in the two subsequent trials. Five of the 12 experimental subjects showed 100% positive responses (mouth opening to the light alone) after only six to ten training sessions. The responses of the other seven are not given. The control animals

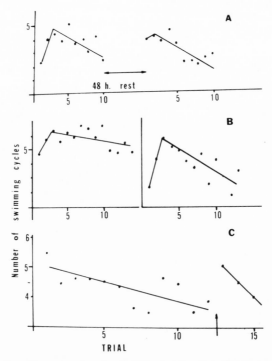

Fig. 15. Habituation of escape-swimming in *Tritonia*. Duration of each response was measured as the total number of ventral–dorsal flexion cycles. Stimulus was a constant small quantity (a few crystals) of salt dropped onto skin. Points represent averages of response of 12 animals run in random sequence. A, Recovery of response duration with time. Left: Ten sequential trials (interstimulus interval 10 min). The rise in response duration seen over first few trials (sensitization) is a common feature of experiments done in this way. It was probably caused by a few (three to five) individuals out of the 12 that did not respond to the first stimuli. After the second or third trial, response duration declined gradually. After the tenth trial came a 48-hr recovery period. Right: On the ten succeeding trials, response level recovered to within 10% of the original. Response decrement occurred at about the same rate. Additional experiments showed that recovery was essentially complete in 3–5 hr. B, Effect of ISI on habituation. Left: ISI 20 min. Decline in duration was to about 80% of original 15 trials. Right: ISI 10 min. Response decreased to approximately 50% of original over 15 trials. Note sensitization of response over first few trials as in A. C, Dishabituation. Points represent average for all *responding* animals in the group. If swimming did not occur, response was not included in total. Initial sensitization was no longer apparent. Response duration declined over 12 trials. Between trials 12 and 13 (arrows), each animal was manually lifted from the water and then replaced in its tank. On trial 13, response level recovered to original level. The decline over trials 13–15 was more rapid than the original. In all experiments, it was unclear whether declines in response duration are best fitted by exponential, linear, or other function. Lines are fitted by eye.

received only the conditioning stimulus for all 12 trials and continued to show only occasional contractions of the tentacles or slight movement.

The mouth-opening response of *Aplysia* consists of several sequential components, and it is not always elicited in its complete form. Also, some of the components are occasionally seen in the absence of any apparent stimulation. Although with experience the various components of the response can be correctly identified, it is nevertheless true that there is a serious hazard that experimenter bias could enter into the interpretation of these responses. Accordingly, they are best done "blind," i.e., by an observer who has no knowledge of the expected results for any particular animal. Recognizing this and other difficulties of interpretation, Jahan-Parvar (personal communication) repeated the work with five specimens of *Aplysia* and trained them according to the procedures of forward, differential, and backward conditioning. Some of these responses have been recorded "blind." A fourth animal received CS and US in random sequences and intervals. The fifth was presented with the CS only during the sessions. The preliminary results of these tests showed that the MOR occurred significantly more often for the forward-conditioned animal than for the backward-conditioned one or for each of the other two control subjects.

Despite the original results and the attempted confirmation, the results of this training scheme are as yet uncertain. Independent attempts (Kupfermann, personal communication) at duplicating them using a larger number of experimental and control animals under conditions which resembled those of the published work as closely as possible have not confirmed Jahan-Parvar's results. All of Kupfermann's observations were made "blind." One can only conclude that either additional "blind" controls are needed in the former experiments or that there are other subtle differences between the two experiments.

Additional work is required to establish the experimental conditions under which the abovementioned differences arise and to establish unequivocally whether or not the MOR of *Aplysia* can be routinely conditioned.

On the positive side, Mpitsos and Davis (personal communication) report that a similar response in *Pleurobranchaea,* a notaspidean, can be readily conditioned if the presentation of food is paired simultaneously with a tactile stimulus. Their experiments were done "blind" and involve a clearly defined biting response. *Pleurobranchaea* normally responds to tactile stimulation of its oral veil by withdrawal. With suitable training, animals made an increasing number of feeding responses to this same stimulus. The CS was several seconds of touching the oral veil with a glass rod. The UCS was the presence of food substance (fresh-ground squid juice) on the glass rod. Training consisted of daily sessions of simultaneous CS–UCS at 30-sec intervals until 20 feeding responses were elicited. Control animals received tactile stimulation only.

After 3–4 days of training and for at least 10 days thereafter, experimental animals showed a statistically significant increase in frequency of feeding responses and a reduction in withdrawal responses and latency to feeding responses, by comparison with controls.

In subsequent experiments, animals that had been trained in this way were shocked for making feeding responses to the CS (20 times daily at 60-sec intervals). After as little as 1 day, subjects ceased feeding in response to the tactile stimulus.

A strong case can be made that an association has been made by the subjects in these experiments. The case could be made even stronger by additional sensitization controls. These seem necessary because the controls in the classical conditioning experiments also showed a significant improvement in performance. In addition, long-term sensitization must be considered .a possibility, since animals making any component of feeding responses during testing received extra stimulation to "determine the maximum strength of the feeding response." Sensitization in other species, e.g., *Physa* (Wells and Wells, 1971) and *Tritonia* (Abraham and Willows, 1971), has been observed to persist for 24–48 hr.

Both Wells and Wells (1971) and Abraham and Willows (1971) found that the pairing of stimuli led to no better performance than when the stimuli were presented randomly in time. Wells and Wells used the same pond snail, *Physa,* as did Thompson (1917) and Sokolov (1959), and attempted to develop a conditioned withdrawal by pairing either the turning off of a light or a thump on the table with electrical shocks (Fig. 16). They conclud-

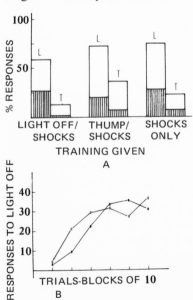

Fig. 16. Sensitization of withdrawal response in *Physa* (Wells and Wells, 1971). Upper: Three groups of ten snails each given 40 trials of either light-off–shock, thump–shock, or shocks only paired training. Total responses to light off (*L*) and thumps (*T*) before (shaded) and after (open bars) training are shown for each group. Training produced neither an increase in responses to the CS by comparison with "shocks only" group nor a significant change in the ratio of light-off responses to responses to thump. Below: Comparison of effects of paired light-off–shocks training (filled circles) with alternating light-off–shocks (open circles). Responses in both groups increased similarly.

ed that pairing has no specific effect on the animal's subsequent responses toward the particular stimulus paired with the shock and that the only effect of the repeated presentation of stimuli is to increase the level of responding in the same way that shocks alone do.

Abraham and Willows (1971) were similarly unsuccessful in establishing a conditioned escape-swimming response in the nudibranch *Tritonia*. As was pointed out earlier, this swimming sequence is normally elicited whenever the animal comes in contact with starfish anywhere on its body. The response consists of a withdrawal sequence of dorsal and ventral swimming cycles and a stereotyped termination. The same response can be elicited by touching the animal anywhere on its skin with a strong solution of soap, detergent, or other surface-active agent, or alternatively with salt solution or crystals. The attempt to demonstrate conditioning was made with 24 animals divided into five groups. Five animals received conditioning trials, five other nonnaive animals (previously used in pilot experiments) received identical conditioning trials, five more received CS trials only, four animals received US trials only, and five animals received both CS and US but in randomized sequence. Three sessions of five trials each were given to each animal each day. The intertrial interval was 1–20 min. The onset of the CS preceded the onset of the US by 10 sec. Since the shortest US–response latency in the US–only group was 5 sec and the CS–US interval was 10 sec, any response occurring in less than 15 sec from the onset of the CS could be supposed to have been triggered by the CS. Accordingly, responses with shorter latencies were scored as conditioned responses. Even after 60 trials, swimming responses to the CS were very infrequent under all experimental conditions and did not vary systematically over time or trials (Table III). It is apparent that the US may have potentiated CS responding, but it is also clear that the randomized group (sensitization controls) showed the greatest proportion of responses to the CS, indicating that if any change in response level can be assumed at all it must be attributed to a generalized sensitization associated with stimulation and not to a temporal association between the two stimuli.

A further experiment, aspects of which can be interpreted either in the context of classical conditioning or as an example of the learning of a passive

Table III. Proportion of Responses to CS

Group	Trials		
	1–30	31–60	61–210
CS	0.00667		
RAND	0.0867	0.0867	
COND—Naive	0.0333	0.0467	0.0551
COND—Nonnaive	0.0333	0.0333	

or inhibitory avoidance behavior, was carried out by Lickey and Berry (1966) and later elaborated on by Lickey (1968). The experiment showed that with training *Aplysia* can learn to reject inappropriate food objects. Twelve *Aplysia* were used in two groups of six. Each group was subjected to three training sessions with 4-day intervals between them. There were three types of stimuli: food trials consisting of touching the lip with a small piece of food (the alga *Ulva*), test trials which resembled the food trials except that the blunt, closed tips of a pair of forceps without any food in them were presented, and "ambiguous" trials which consisted of a presentation of forceps in which the food was held well up in the jaws of the instrument. In this last-mentioned case, stimulation came initially from forceps contact, and if these were ingested the food could then be obtained. Each session consisted initially of five food trials, which served to facilitate food ingestion for the remaining trials, and then either 30 test–food trial pairs or 30 "ambiguous" trials. A session in which test and food trials were alternated was called a "test" session and the others were called "ambiguous" sessions. Responses were scored in terms of whether the animal ingested the food or forceps or whether it rejected them. The animals never rejected food, and in every case ingested it within 15–20 sec. Overall, animals rejected the forceps plus food in the "ambiguous" trials in about 40% of cases and rejected the forceps in the test trials in about 87% of cases, indicating that the forceps served as an effective negative stimulus for the feeding response.

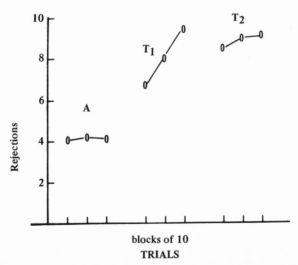

Fig. 17. Learned discrimination between food and nonfood objects in *Aplysia* (Lickey, 1968). Rejection of nonfood object per ten trials remained low over 30 "ambiguous" trials (*A*) but increased when "training" was given (*T*₁) and remained high after 4 days (*T*₂).

The principal result was that during the first session in which food and test trials alternated the animals all showed a marked increase in their tendency to reject forceps when they were presented without food (Fig. 17). On the average, during the first block of ten test trials the number of rejections was about 6.7, whereas during the last ten trials of that same 30-trial sequence the number of rejections had increased to nearly 9.5. Furthermore, this increased number of rejections of the inappropriate stimulus after 30 trials remained high for as long as 4 days in this experiment and 9 days in the earlier experiment (Lickey and Berry, 1966). The number of rejection responses in the "ambiguous" sessions did not increase over the 30 presentations. Another finding of potential interest was that the latency to ingestion in the "ambiguous" session tended to increase, suggesting that the "ambiguous" stimulus tended to become less and less attractive as a food object with successive presentations.

As the author points out, there is an alternative explanation to the suggestion that the animal is learning an inhibitory or passive avoidance behavior. The change may in fact be a habituation of the ingestion response or a potentiation of the competing responses to the forceps. The habituation would not necessarily be considered associative learning, while the acquisition of the potentiation for the competing response would. A conclusion in the context of classical conditioning may be drawn from the finding that the latency to ingestion in the "ambiguous" trials increased significantly over the three blocks of ten. If the forceps presentation can be considered as a conditioning stimulus and the delivery of the food which was held higher in the forceps as an unconditioned stimulus for the ingestion response, then one might look for a reduction in the latency to ingestion as a sign that the animal was associating the two stimuli. The fact that, if anything, this latency became longer must be considered evidence against the formation of a conditioned reflex.

b. Operant Conditioning. Lee (1969) showed that animals could be made to enter a particular area of their aquarium at a higher rate than their yoked controls provided that correct performance was reinforced by raising the water level from 1 to 3.5 inches. (It has since been shown by Downey and Jahan-Parvar, personal communication, that the reinforcing effect of the water change may reside in the reduction in temperature associated with an increase in water depth.) The aquaria used by Lee were rectangular, with either a small vertical notch in one end or a ledge in one corner under which the animal could place its head and break a light beam and thereby activate an external circuit controlling a pump. The pump would then raise the water level. Using eight subjects with yoked controls, he found that all the experimental animals responded more consistently than their yoked controls. Perhaps even more convincing was the finding that if experimental and control animals were periodically removed from the aquarium and replaced facing away from the goal, all the experimental subjects showed shorter latencies for the response than did their yoked controls. An interpretation of this finding is that the animals had learned that movement into a particular region of the aquarium led to an increase in water depth.

In later experiments, Lee (1970) shaped the behavior of *Aplysia* by again using contingent water-level changes as a reinforcer. The apparatus consisted of a pair of identical chambers sharing water input and output (Fig. 18). At one end of each chamber was a vertical slot, 1 inch on each side, which ran to the top of the aquarium wall. Photocells were placed in successively higher locations in the slot so that the animal had to reach up to break the beam. Reinforcement was provided for successively higher movement up the slot. In this experiment, the duration of the reinforcement equaled the duration of the required response. When performance at a given level improved, the contingency height in the slot for a water-level increase was raised. The experiment was continued for as long as the animals or the

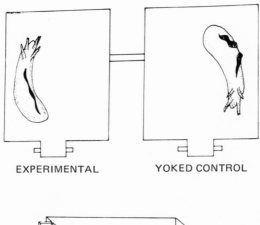

EXPERIMENTAL   YOKED CONTROL

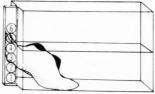

Fig. 18. Apparatus used in operant conditioning of *Aplysia* (Lee, 1970). Upper: Water levels in experimental and control chambers rise together contingent on movement of experimental animal into specified region of aquarium. Lower: Photocells (*1–6*) at successively higher levels in notch at end of aquarium activate pump to raise water level in experiments to shape behavior.

apparatus would permit, a period ranging from 29 through 192 hr for different pairs of animals. In all cases, the number of hours spent by the experimental animals at the fifth level in the slot was substantially greater than that spent by the control animals (Table IV). In one instance, the experimental and yoked animals were reversed and their behaviors changed appropriately. Four out of the six experimental animals increasingly improved their performance over the first several hours of training.

A similar experiment in which the animal was required to pass into the higher level in order to be reinforced with a fixed duration of increased water level had less consistent results. Five pairs of animals (experimental plus yoked controls) were used, with reinforcement (i.e., raised water level) durations ranging between 4 and 16 min. The number of half-hour periods with at least one reinforcement under the position 5 contingency was recorded for each animal over the entire training period. Three of the five pairs showed substantially higher response levels for the experimental over the control, and in one case the shaping of the behavior under higher and higher requirements was particularly striking (Fig. 19). Although the data for all of the control subjects in this experiment were not provided, it is clear from the results for two of the subjects that the probability of a spontaneous response at the fifth level in the slot on the basis of no reward is very low indeed.

Wells and Wells (1971), alongside their classical conditioning experiments reported above, tried to operantly condition snails *(Physa)* to *not* cross a light–dark, smooth–rough barrier in a watchglass filled with water. Each time an animal crossed the barrier it was given an electric shock, and at the same time its yoked control was shocked regardless of its position or

Table IV

Animal	Duration of training (hr)	Hours of position 5 response
28	29	2.45
29		0.00
30	192	2.09
31		0.54
31	95	2.90
30		0.57
33	113	20.47
34		4.10
36	30	5.25
37		1.41
38	46	4.49
39		1.15
40	36	8.59
41		0.94

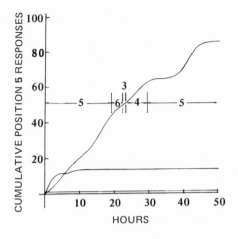

Fig. 19. Improved performance of one *Aplysia* with successively higher contingency levels (Lee, 1970). Responses at position 5 in notch rose continuously with changing contingencies. Required response levels are shown for each stage of training. Two control subjects run with water continuously down showed consistently low response levels.

activity. Barrier-crossing activity of an unstimulated group of snails was also monitored. Each group consisted of 16 animals. When the number of barrier crossings per 5-min interval is plotted against time, there is no significant difference between any pair of the three groups. Both the experimental and the control animals gradually reduced their movements in the dish. The conclusions drawn by these authors from both their classical and operant conditioning experiments is that *"Physa* does not possess nervous machinery for the separation of significant series of events on a basis of their sequence in time. The animals, in short, can be sensitized but not conditioned."

c. Maze Learning. Thompson's (1917) attempts to train *Physa* to turn in one direction in a submerged U-shaped labyrinth in order to reach the surface were generally unsuccessful. She took advantage of the fact that if the air is gently squeezed from the gill cavity and the snail is dropped to the bottom of the aquarium, it immediately climbs up by the first available route to the surface to refill with air. She therefore made a U-shaped route at the top of a short runway so that a choice had to be made between turning left (leading to a dead end, with no access to air) or right (leading up to the surface, where the snail could refill its gill cavity); 888 trials on this labyrinth led to 59% incorrect choices and the conclusion that the animal did not acquire an obvious left–right preference when it was reinforced by this method. She then modified the apparatus by roughening a strip in the path on the incorrect choices. Finally, in an apparatus similar to the previous one, if the animal crossed the rough spot in the pathway she tapped it gently on one of its tentacles with a fine brush and noted in how many cases the animal turned around and went back via the correct route to the surface. She made a few trials with the brush alone to see if the brush stimulus served to drive a turning-around response. It apparently did not. However, in 930 tests 33 % of the animals took the short arm and received the punishment as well as the brush stimulus on the tentacles, despite the fact that it led to a shock at the end, with no air. Only 51 % of the animals took the long arm directly. All of the rest (16 %) turned back to go the correct route. She concluded that these 16 % of subjects had learned by associating the brush with an incorrect choice. Without more and better controls for the effect of mechanical stimulation on the tentacles as an aversive agent leading to a reversal of direction, this conclusion cannot be drawn confidently.

Thompson's experiments, just described, and Yerkes' (1912) work with earthworms stimulated similar attempts by Garth (1924) and Garth and Mitchell (1926) using the snail *Rumina*. The length of time taken to reach a goal box in a glass T-maze was measured in groups of five trials. Over 102 trials, the average time in the maze was 856 sec, and for the last five trials the comparable average was 316 sec. The number of errors per five trials declined from four in the first sequence to zero in the last. In the second report, a more complete description of the apparatus and the results was given. The T had an 18-cm-long leg and two 15-cm arms. It wasm ade from window glass and was washed between each trial to prevent the snails from tracking on slime trails. The snails were known to be negatively phototropic. Accordingly, their activity level was maintained high by exposing them to an electric light from above. The right arm of the maze was always the correct one. The end of the arm was darkened so that when the animal made a correct choice and reached the end of the arm it was reinforced by being permitted to remain in the dark box for a few moments. The time was recorded from the moment the snail started up the stem of the T until it reached the dark box. From one to six trials were given each day, depending on the activity level of the snail. The maze performance of the animal was monitored daily for nearly 3 months. Over that period, the time taken to reach the goal box declined from approximately 1000 to approximately 300 sec. It is noteworthy that the response time became much more constant toward the end of training and furthermore that the number of complete failures, i.e., failure of the animal to move at all, declined rapidly over the first 30 trials (Fig. 20). With 30 days of rest, there was a substantially increased time in the maze, but nevertheless it was still much shorter than at the beginning of training period. A second line of evidence for learning cited by these authors is the decline in the number of errors made per five trials.

An examination of the tracks followed by ten different snails in the maze shows that generally they were extremely tortuous and it was only rarely that an animal made a direct approach to the darkened goal box. The authors cite one individual, however, that made increasingly direct approaches to the box and rarely made an incorrect turn. However, its time in the maze did

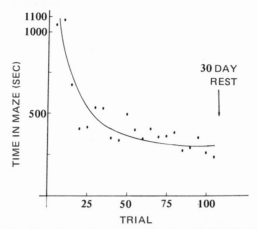

Fig. 20. Performance in a T-maze in *Rumina* (Garth and Mitchell, 1926). Average time in the maze in one subject declined from over 1000 sec to about 300 sec after 100 trials. A 30-day rest was followed by increase in time in maze. Each point represents average of five trials.

not decline appreciably. The significance of this finding is difficult to assess, particularly in view of the fact that it is a result from a single snail from an unknown total number of subjects.

In the context of Fischel's (1931) experiments, the interpretation of the behavior of this one snail can be seen in a different light. His results with *Lymnaea* and *Ampullaria,* both freshwater snails, are particularly noteworthy, since they include attempts to use natural environmental conditions in the learning experiments. For instance, *Lymnaea* was placed in a runway terminated by a row of pebbles at one end. The snail was required to make an abrupt 90° turn to the right when it encountered this obstacle. When a correct turn was made, the animal could move from the aquarium floor onto an earth-covered runway which presumably more closely matched its natural environment. Turning to the left was impossible, owing to the presence of a physical barrier. If it failed to turn, it was coaxed around the corner with pokes from a soft, fine brush. Fischel found, however, that even with 100 trials the total number of turns, including both abrupt 90° and only partial turns, did not increase substantially nor did the number of 90° turns increase over the number of partial turns. Similar attempts to train *Ampullaria* in the T-maze were not successful.

On the other hand, *Ampullaria* did improve its performance in a somewhat different circumstance. A long straight glass runway was placed vertically in an aquarium, and at a point 3 cm from the top a strip of jagged tin was attached to the inset. Each time a snail crawled up to and encountered the tin, it was poked with a pin. After 3 weeks of this training, the animals withdrew spontaneously whenever they encountered the tin. Performance over the 3 weeks improved continuously, and responses are shown in Fig. 21. No attempt was made to determine whether the response was specific to the stimulus or instead caused by a generalized sensitization evoked by the repeated pinpricks. Under these circumstances, it would not be surprising to find that sensitization had occurred.

In another set of experiments, *Ampullaria* was made to crawl through a runway to a choice point. Three alternative channels were available, one of which was blocked by shock-delivering electrodes and the other two were open. The snails did not learn to avoid the shock electrodes successfully. However, if a snail entered the same arm in the maze several times in succession (either on the basis of chance alone or due to a dominant tendency to turn one way), it sometimes stuck with this particular pattern of movement. By the next day, this kind of repetition ceased. It may be that the animal was training itself by repetitive performance of the same sequence of movements in the maze by kinesthetic cues. The behavior was probably not the result of association of the path taken with reinforcement, since the choice often led to shocks.

A more likely interpretation of the apparent improvement in maze performance in the experiments of both Garth and Mitchell (1926) and Fischel (1931) is that snails were following odor trails laid down in earlier trials. In neither experiment is it clear that the maze cleaning between trials was sufficiently thorough to obliterate remnants of mucous from previous trials. Successful performance in both these experiments required only that animals

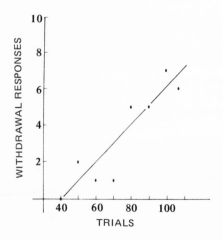

Fig. 21. Withdrawal responses upon contacting jagged metal surface over trials (Fischel, 1931). Whenever an individual *Ampullaria* came to tin on floor of runway, its tail was pricked with a pin, eliciting withdrawal. Withdrawal in response to contact with tin (before pinprick) increased steadily over 100 trials.

repeatedly follow approximately the same track. Both Cook *et al.* (1969) and Crisp (1969) have shown that limpet and snail odor trails may be extremely persistent. Likewise, Hall's finding (described above) that *Littorina* recognizes the polarity of a mucous trail is likely to be relevant.

d. Homing. Both limpets and numerous marine prosobranchs show clear evidence of a capability to locate either their home area or, in other cases, their egg masses when displaced from distances of several centimeters to as much as 10 m (Hewatt, 1940; Arey and Crozier, 1921; Davis, 1895; Fischer, 1898; reviews by Thorpe, 1963, and Wells, 1965). Many gastropods return regularly to a restricted location on a rock in the intertidal zone, which becomes their home base over long periods of time. Often this home base is a protected flat area on the rocks or a region where cracks and fissures in the rocks provide a natural indentation into which the animal may fit its shell fairly snugly. It is typical in some species for individuals to forage at distances of many centimeters from their home bases. Commonly, they return on a regular schedule linked to the tidal movements.

Some pulmonate slugs and snails exhibit marked homing capabilities, too (*Helix*—Edelstam and Palmer, 1950; *Limax*—Gelperin, personal communication; *Agriolimax*—Newell, 1966). Ordinarily, these terrestrial forms remain hidden beneath particular rocks or other protective objects except during the night or times of high humidity. They may then be found foraging at distances of several meters. Apparently, the capability to return does not require vision, since total darkness does not impair homing. Slime trail following has been discounted as the homing mechanism because homing is not seriously impaired after mucous trails are brushed or scoured off the substrate. This factor may require reconsideration in light of the findings mentioned above that slime trails may be extremely persistent.

A question naturally arises (1) whether homing depends on an internalized memory of the route or of the sequence of physical or chemical cues along the numerous routes toward home or (2) whether it involves following a slime trail, a chemical gradient, a physical path (e.g., cracks or ledges in the rocks), or other paths which inevitably lead to the home base. Another possibility is that homing may be an essentially random process. This would be possible if movements around the home base were restricted by natural obstacles (cracks, rocks, or ledges) to a relatively small area. Tidal activity might then merely stimulate unguided locomotion, which would continue until the scar in the rock was encountered by chance.

Cook (1969) showed that the location of the horizon or other cues associated with the shoreline or plane of polarized light are not essential to homing in the pulmonate limpet *Siphonaria*. Individuals were found to home successfully despite rotation of their rocks through 90° or 180°. Neither was homing a random process. Provided that an animal had been observed in an area at some time in the recent past, it could normally return home directly, with little or no random wandering. Return was less likely and indirect if the snail was placed in a location not recently visited. It was concluded that homing in this species depends either on olfactory cues associated with previous trails or on a "memory" of these paths and the nearby locality.

Several members of the prosobranch family Calyptraeidae, including *Crepidula* and *Calyptraea,* have well-developed egg-brooding behavior (Fretter and Graham, 1962). Also, the large marine prosobranch *Fusitriton* remains directly on top of its mass of egg capsules and ejects intruders either by pushing them off or by squirting a noxious fluid at them from its proboscis (C. Eaton, personal communication). Normally, *Fusitriton* makes only short excursions away from the egg capsules during the period of brooding, but on its return it apparently has no difficulty relocating its own from among the several other individuals' masses that are packed together on the same rock surface. Surprisingly, however, it is also capable of returning to its own eggs from distances of as great as 10 m in any direction. The precise nature of the cues used to home under these conditions is not known, although both chemosensation and vision seem unlikely in view of the water currents and poor visibility which exist at times in the environment. This remarkable capability invites further study and may prove to be of utility in assessing the role that learning plays in homing behavior.

Among the various alternative explanations that have been suggested for homing behavior in gastropods, the two that seem best supported by experimental evidence are (1) movement down a chemical gradient, the focus being a slime trail or the home itself, or (2) memory of physical–chemical aspects of the local environment. It is also possible, of course, that a combination of these two enables gastropods to return to their homes.

Some carnivorous snails such as *Natica* and *Nucella* feed by boring holes, with the aid of the radula and secreted solvent solutions, in the shells of bivalve mollusks and barnacles. There are both good and bad locations for the placement of such holes, since the shells of these prey animals vary in thickness and the underlying tissues vary in edibility and nutritional value. Efficient boring performance has been found to improve as individuals mature (Verlaine, 1936). In another case, Fischer-Piette (1935) reported that when their barnacle bed became overrun by mussels, a colony of the whelk *Nucella* took to eating the mussels. Furthermore, skill at choosing an appropriate location for the hole in the shell improved markedly over time and, astonishingly, remained high in succeeding generations of snails. Interpretations of the noted improved performances in feeding behaviors not requiring learning are possible. One, for instance, is that the predator may have changed the location for hole boring as its own size increased because with larger size it was better physically adapted for making holes in particular locations which also happened to be better suited in terms of obtaining food. This, of course, does not explain the apparent transmission of an acquired characteristic. Both homing and boring activities have obvious adaptive values and almost certainly represent an excellent opportunity for further study with emphasis on isolation of possible learning components.

B. Neurophysiological Studies of Learning

Apart from demonstrating that aspects of gastropod behavior can habituate, sensitize, and perhaps be conditioned, the behavioral studies already described permit the inference that most of the behavioral changes result from alterations in the activity of central neurons, as opposed to sensory, peripheral nerve net, or neuromuscular elements. This naturally leads to the question of the precise neural location and mechanism of the observed behavioral change. One might expect that the most logical sequence in which to pose these questions experimentally might be in the order given, namely, first determine the specific neuronal pathway that controls the learned behavior and then study the underlying mechanism. Curiously, the approach taken has been the reverse. Until very recently, randomly chosen neurons were stimulated electrically in temporal and spatial patterns that formally resembled the patterns of stimulation used in behavioral experiments. Later, when specific neurons and pathways for certain modifiable behaviors became known, the insight gained from the earlier studies of arbitrarily chosen neurons became the basis for an explanation of the cellular-level mechanism of the behavioral plasticity.

1. Neural Analogies for Habituation

In part motivated by an earlier finding of Hughes and Tauc (1963), Bruner and Tauc (1964, 1965a,b) found that electrical changes formally resembling habituation occurred in several nerve cells of *Aplysia* in response to nerve or sensory stimulation. Their preparation was made by removing the posterior body and visceral mass, keeping the nerves from the various sensory structures of the head and from the parietovisceral ganglion to the periesophageal ganglia intact. They arranged the preparation (Fig. 22) so that they could record from the giant cells in the left pleural ganglion and the abdominal ganglion while stimulating the remaining sensory structures of the head with water drops. There were two parallel observations. First, water dropped from a 30-cm height onto the head every 10 sec produced a reflexive withdrawal of the tentacle and simultaneously a 10–15-mv composed EPSP in the two giant cells, located in the left pleural and parietovisceral ganglia, respectively. With repeated stimulation, both the tentacle reflex and the compound EPSP declined in amplitude. If the water drops were

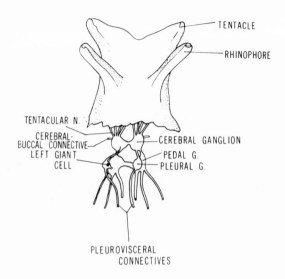

Fig. 22. Intact head preparation of *Aplysia* (after Bruner and Tauc, 1965*b*). The head, including tentacles, rhinophores, and ganglia, was removed and pinned out. Water-drop stimuli applied to the head region elicited tentacle withdrawal which habituated upon repetition. Likewise, water-drop stimuli evoked compound EPSPs in the giant cell of the left pleural ganglinon which also decreased in amplitude.

terminated for a few minutes, both the reflex and the compound EPSP recovered. Upon recommencing stimulation, the rate of decline in amplitude increased by comparison with the first series; i.e., rehabituation was more rapid. Finally, they noted that a novel stimulus (a scratch applied to the skin or in other experiments an electrical stimulus to the pedal or parapodal nerves) delivered between two of the water drops led to recovery of both electrical and behavioral responses nearly to the original level. There is no indication that either of these cells, R2 or the left pleural giant cell, drives or participates in the tentacle reflex directly, but the suggestion is clear that the decremental phenomona may have a common causal basis. The same decline in EPSP amplitude can be noted in numerous other cells in the ganglion.

In an isolated nervous system preparation, Bruner and Tauc were able to show that the reduction in compound EPSP amplitude was probably not caused by reduction in the number of unitary EPSPs making up the compound response. It was more likely to be the result of a decline in the amplitudes of the unitary EPSPs. By reducing the amplitude of stimulation to one of the cerebral nerves, they could elicit a unitary EPSP having an amplitude of 4–6 mv, and the amplitude of this unitary response itself declined with repeated stimulation. In addition, it recovered with rest and could be restored by interposing 3 sec of repetitive stimulation at 10/sec via another cerebral nerve or by stimulating at a higher frequency on the same nerve (Fig. 23).

On the basis of these experiments, Bruner and Tauc suggested a mechanism and a location for the decrement in amplitude of the EPSP. The alteration was apparently *not* taking place in the postsynaptic membrane or neuron, since direct measurements showed no change in the membrane conductance of the postsynaptic cell during "habituation" of the EPSP. Apparently, the postsynaptic cell was responding in its usual way to a *changed* presynaptic signal. This conclusion is further buttressed by the observation that the EPSP recovered after a brief tetanization through the same input (i.e., it "dishabituated"). This argues against desensitization of the subsynaptic receptor sites as the responsible mechanism for the EPSP decline. Presynaptic inhibition, i.e., an inhibition of the presynaptic cell itself, is unlikely, because in other experiments Bruner and Tauc showed that such inhibition is reduced and *less* effective with repetitive stimulation. The conclusion drawn is that during habituation the individual EPSPs in the pathway are reduced because of a reduction in the amount of transmitter released from the presynaptic terminals. This phenomenon is probably caused by depletion of the transmitter in the terminals and/or by its slower mobilization. This clearly stated hypothesis has found additional support from experiments described in succeeding sections involving other neurons and reflexes in *Aplysia*. The question of the mechanism underlying the reduction of transmitter release still remains.

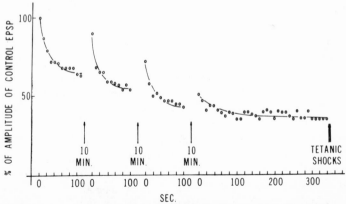

Fig. 23. Decrement in unitary EPSP in left giant cell with repetitive stimuli to a cerebral nerve of *Aplysia*. Isolated nervous system preparation (Bruner and Tauc, 1965*b*). Decrement was interrupted by 10-min rest periods, after which partial recovery and rehabituation were observed. Recovery was less after successive rests. Tetanic shocks at 10/sec to another cerebral nerve produced nearly complete recovery.

2. Neural Analogies for Classical Conditioning

Whenever classical conditioning is demonstrated in behavior, one can assume that underlying changes take place in the central neurons controlling the behavior. In behavioral terms, classical conditioning normally requires that response occur with increased vigor or reduced latency or to a stimulus to which it was previously insensitive. Since an improvement in performance might result from an increase in general sensitivity to many or all kinds of stimulation caused by the arousing effect of the experimental stimulation, a distinction is made between conditioning that is specific to the *pairing* of the CS and US in time and that which is not. It is on this basis that classical conditioning and sensitization are separated. The most naive interpretation of the neural basis of classical conditioning is that a stimulus which normally evokes a response in a motor pathway comes to potentiate the efficacy of stimuli arriving on another sensory channel. This, it is assumed, increases the likelihood that this new input will elicit the same response. For the potentiation to be classical conditioning, it must be contingent on the close temporal contiguity of the two stimuli.

Despite the several attempts described above to demonstrate classical conditioning in gastropod behavior, few unequivocal successes can be cited. Unfortunately also, despite the recent impressive progress that has been made in determining the functional roles of central neurons in gastropods (see below), the detailed neural circuitry for any behavior which is likely to be classically conditioned has not yet been worked out. The most clear-cut progress has been made in analyzing the neural apparatus controlling gill withdrawal (see below) and escape swimming (Willows *et al.*, 1973*a,b*), while the kinds of behaviors which have shown the most promise for classical conditioning have been related to feeding activities. However, notwithstanding this uneven progress, attempts have been made to analyze a classical conditioning paradigm in the isolated ganglia of *Aplysia*. Results indicate that sensitization may have a readily identifiable neural correlate and that specific classical conditioning may eventually be understood in terms of a neural mechanism which derives from the one shown to apply to the non-specific form.

Intracellularly recorded EPSPs from neurons of the abdominal ganglion of *Aplysia* increase in amplitude for several minutes or longer as a result of synaptic stimulation through a separate pathway in a sequence of trials which formally resembles the classical conditioning scheme (Kandel and Tauc, 1965*a,b*; von Baumgarten and Jahan-Parvar, 1967*a,b*; Jahan-Parvar and von Baumgarten, 1967; von Baumgarten and Hukuhara, 1969; Tauc and Epstein, 1967). In most cases, these experiments follow a similar format. One neuron in the ganglion is penetrated with an intracellular micropipette, and its response to stimulation via two different nerve trunks is monitored.

A single shock (analogous to the CS) is applied to one nerve trunk, a delay of the order of 200–700 msec follows, and then a short train of stimuli, usually stronger in amplitude and at about 10/sec for 0.5–1.0 sec, is delivered to another nerve trunk (the UCS). At the outset of the experiment, the response of the cell to the CS, called the "test" stimulus, is a small, compound EPSP. The response of the same cell to the UCS, called the "priming" stimulus, is either a sequence of much larger EPSPs or a short volley of impulses. In a successful experiment, after pairing the two stimuli every 10–20 sec for a period of 5–20 min, the amplitude of the response to the test stimulus increases severalfold (Fig. 24). This phenomenon is called heterosynaptic facilitation (HSF) and is considered to be a possible neural analogue of classical conditioning. The amplitude of the EPSP increases from 40% to over 150% over its control amplitude during a period of 2–5 min and then, with the termination of paired stimulation, declines with a time course of a few minutes. Both Bruner and Tauc (1965b) and Jahan-Parvar and von Baumgarten (1967) report that in some neurons HSF is preceded by a brief period of depression of the EPSP, a phenomenon described as heterosynaptic inhibition (HSI) by the former authors.

The time relationship between the two stimuli is critical to the efficacy of this experiment as an analogue to classical conditioning. For instance, in a behavioral experiment with classical conditioning, if the two stimuli were applied either in random sequence, or in reverse order, or even singly, and if under any of these circumstances the response to the conditioning stimulus alone subsequently increased, the behavior would not be said to be classically conditioned but instead to be sensitized or quasiconditioned. A neurophysiological parallel to this kind of behavioral control experiment done by omitting the test shocks was performed directly in the case of three neurons

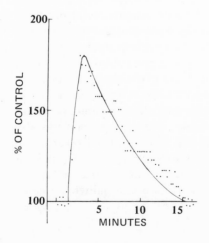

Fig. 24. Time course of heterosynaptic facilitation of a presumed unitary EPSP in an unidentified neuron of the parietovisceral ganglion of *Aplysia* (Kandel and Tauc, 1965a). Four paired trials ("test" stimulus to right connective, "priming" stimulus to siphon nerve) produced increase of 180% in amplitude of EPSP over 2–3 min which decayed during the next 10–12 min.

by Kandel and Tauc (1965a) and another type of control was used in two other neurons. In these five cases, HSF only occurred with paired stimulation. Overall, about 80 cells were tested in the study and 15 showed HSF, and in the abovementioned five cases out of the 15 the increase may have been specific to the pairing of stimuli. HSF in the other ten neurons was not checked for specificity. Attempts by Kupfermann and Kandel (reported in Kandel and Spencer, 1968) to duplicate these controls for specificity were not successful. Jahan-Parvar and von Baumgarten (1967) studied a comparable number of neurons in the parietovisceral and left pleural ganglia of *Aplysia* and found that between 10% and 15% of the cells tested showed HSF but that none of these cases were entirely dependent on pairing of the stimuli. Although in no case could HSF be shown to absolutely be dependent on pairing of test and priming stimuli, the extent of HSF was greater when paired stimuli were used. As well, the strength of stimulation and choice of nerve trunks markedly influenced the intensity of HSF produced. Weaker stimuli than those used by Kandel and Tauc (1965a,b) were more effective, and certain nerve trunks established better HSF than others. In a more recent report, von Baumgarten and Hukuhara (1969) found that HSF depended strongly on the interval between test and priming stimuli, with most pronounced facilitation of the EPSP at an interval of 350 msec. This finding is consistent with the observation that in mammals classical conditioning is best demonstrated when CS and US are paired at intervals of this magnitude. On the other hand, these same authors found little or no HSF when the stimuli were *simultaneously* paired or paired with a longer interval, and it was with just such intervals that behavioral classical conditioning has been reported in gastropods (e.g., Thompson, 1917; Sokolov, 1959; Jahan-Parvar, 1970; Mpitsos and Davis, personal communication).

In either case, specific or nonspecific, HSF is clearly a fundamental form of neural plasticity whose mechanism is of unusual theoretical and practical interest. It is most likely not postsynaptic in origin, because on the one hand there is no change in the conductance of the postsynaptic membrane seen during priming stimulation and on the other hand spikes initiated directly in the cell are not an adequate substitute for an orthodroic priming stimulation producing HSF (Kandel and Tauc, 1965b; Wurtz et al., 1967). Disinhibition as a mechanism for the increase in the amplitude of the individual EPSPs is not considered likely, because the configuration of the compound EPSPs remained essentially the same at all amplitudes. The phenomenon was also studied using a unitary EPSP, and again it was found that the depolarization increased by as much as 100% over control values during HSF. The conclusion that can be most reasonably drawn is that a change is taking place at a presynaptic level. Most likely, an increased amount of transmitter is released for each test stimulus.

A mechanism based on post-tetanic potentiation (PTP) is also possible, in principle, since the train of shocks in the "priming" pathway could indirectly elicit activity in the "test" pathway. This latter activity itself might then facilitate subsequent activation via the "test" pathway. Tauc and Epstein (1967) have shown that this alternative is unlikely on at least two grounds. They selectively and temporarily blocked conduction from the "priming" pathway to the cell body of the relevant neuron of the "test" pathway (thereby eliminating repetitive firing in the "test" neuron as a source of PTP) and found that HSF in neuron R2 was unaffected. In addition, low-temperature and Na^+-free medium (which both abolish PTP) had no effect on HSF.

No correlated behavioral–neurophysiological studies of HSF have been made in *Aplysia* or any other gastropod. Until the specific functional roles played by neurons showing HSF are known, the significance of the phenomenon cannot be fairly assessed. However, at least two experiments, in which intact animals were used, suggest that classical conditioning in behavior and HSF may have a common basis. Kandel and Tauc (1965*b*) found, for instance, that if they monitored the electrical activity of the right giant cell (R2) extracellularly in the right connective, spiking activity developed after pairing a "test" input (a shock to the cut left connective) with "priming" stimulation (stroking the region of the siphon with blunt forceps). Presumably, the appearance of spikes coincided with sufficient growth of HSF to cause a suprathreshold depolarization of R2 when the "test" stimulus occurred. The delay to the volley of spikes produced by the "test" stimulus increased as the facilitation declined.

It was mentioned earlier that Jahan-Parvar (1970) has provided evidence for the classical conditioning of the mouth-opening response of *Aplysia,* using a flash of light as a CS and skin contact with food as the US. More recently, the same author (1971) reported that a number of neurons in the left pleural ganglion receive synaptic inputs from both visual and chemosensory pathways, i.e., the same stimulus modes used in his earlier behavioral experiments. They also show relatively long-lasting facilitation when stimulated by electrical or physiological means in the heterosynaptic facilitation paradigm. Although the specific role of any of these cells in the control of the mouth-opening response or any other behavior is not yet known, the result is consistent with the interpretation that HSF may underlie some forms of classically conditioned behavior.

The most simple neural models for HSF include priming nerve synaptic inputs onto the terminals of "test" fibers (see reviews of Kandel and Spencer, 1968; Kupfermann and Pinsker, 1969, 1970). Alternatively, the "priming" input may influence the test inputs via one or more interneurons, having the same net effect (Fig. 25). The transmitter released by these presynaptic

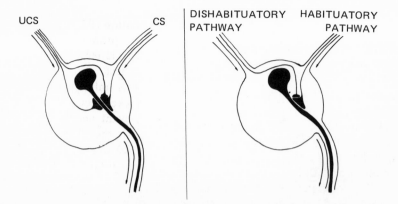

Fig. 25. Diagrammatic representation of simplified neural circuits which may be responsible for heterosynaptic facilitation (left) and habituation–dishabituation (right). In the case of HSF, "test" stimuli (CS) arrive on one synaptic pathway to the recorded neuron. The "priming" stimulation may affect the recorded neuron both directly via synaptic contacts and indirectly by a presynaptic, long-lasting excitatory influence on the terminals of the "test" pathway. Habituation (right) caused by a decline in the efficacy of one synapse may be counteracted by the presynaptic influence from another nerve (the dishabituatory pathway). HSF and dishabituation may therefore be explained by similar or identical synaptic mechanisms.

terminals would establish (by an unknown mechanism) a long-lasting increase in the amount of transmitter released from the terminals for a given terminal depolarization.

In the context of such presynaptic models, von Baumgarten and Jahan-Parvar (1967*b*) propose that the distinction between specific and nonspecific HSF may be understood in terms of the number of interconnections between the priming pathway and the test pathway. They suggest that the distinction made between the two cases, i.e., specific and nonspecific HSF, is in fact an artificial one, since the two may represent extremes along a continuum from the case where the priming nerve makes no excitatory synaptic connections onto the test units to the other extreme where the priming nerve contains numerous axons which excite the test units. By their argument, nonspecific HSF (and therefore sensitization) would come about because activity on the priming nerves at any time would have numerous pathways by which to produce a relatively long-lasting change in the test units. At the other extreme, the case of pathways producing specific HSF, the two pathways could only interact weakly through a small number of collaterals through the priming neurons; simultaneous or near simultaneous excitation of the two would be required in order to fire the test units. The more common circumstance seen in their experiments was a transitional type of response midway between these two extremes.

a. Patterned Firing and HSF. Kristan and Gerstein (1970) and Kristan (1971) looked for changes in the pattern of firing of neurons relative to other neurons under experimental circumstances identical to those producing HSF. They developed a clever technique for the display of stimulus-related sequences of nerve impulses from two or three neurons simultaneously, which draws attention to any stimulus-related and/or synchronous firing tendencies (the joint peristimulus–time histogram). They found that either increasing or decreasing synchrony in the firing of different neurons could occur contingent on pairing of "test" and "priming" stimulation and that changes of both kinds occurred which lasted from 5 to 30 min. They did not attempt to ascertain whether the timing or order of the presentation of the priming and test stimuli was important. They did find, however, that in at least one-third of the over 70 cell pairs that showed alterations in the likelihood of synchronous firing during paired presentations, the effects did not appear in response to the priming stimulus alone. The relationship that increased or decreased firing synchrony bears to HSF or classical conditioning is as yet unclear.

3. Cellular Studies of Operant Conditioning

It is a common observation that gastropod neurons produce recurrent bursts of impulses, in some cases continuing over long periods of time. In particular, certain identified cells of the parietovisceral ganglion of *Aplysia* produce such bursts and have been studied extensively. The oscillatory membrane potential changes underlying bursting are considered to be dependent on endogenous changes in either ionic conductance or electrogenic pump activity (Strumwasser, 1967a; Wachtel, personal communication). Frazier *et al.* (1965) and Pinsker and Kandel (1967) have shown that increases or decreases in interburst intervals lasting for several minutes can be produced by stimulation of such neurons through mixed excitatory and inhibitory synaptic pathways in selected phase relationships with the ongoing bursts. For instance, brief shock trains of moderate intensity delivered to a nerve trunk during the silent period between bursts caused a prolongation of the interburst interval. Stimulation consistently delivered at the time of burst onset produced a shorter interburst interval. The alterations in firing patterns elicited in these ways persisted from 1 to 30 min after termination of stimulation. Since no prominent change in synaptic input could be seen to be directly driving the altered bursting rhythms, the conclusion was reached that the modification probably involved membrane processes endogenous to the neuron rather than extrinsic factors.

Nothing was known then about the functional roles played by the neurons in question in terms of behavior, and to date this information is still lacking. This uncertainty coupled with the fact that a legitimate question has been raised about the validity of such bursting as representative of a normal *in vivo* phenomenon (see below) suggests that further work to clarify these points will have to be done before the relevance of the operant conditioning paradigm can be assessed.

4. Entrainment

Strumwasser (1967*b*) and Kupfermann (1968) have demonstrated that *Aplysia*'s locomotor behavior undergoes a circadian rhythm. By recording the behavioral activity of *Aplysia*, either by using time-lapse cinematography or by measuring animal movements by the mechanical displacement of vertical rods attached to microswitches, these authors have shown that on a schedule of 12 hr of dark and 12 hr of light, locomotor activity occurs largely during the periods of light (Fig. 26). Apparently, the rhythm in locomotor behavior does not depend for its expression entirely on sensory stimulation from light, since the same animal when placed in continuous dark shows recognizable peaks in activity phased with the projected periods of light. Provided that there were no other cues available to the animals about the time of day (the animals were not fed during the periods of these experiments to eliminate time of feeding as a cue), the simplest interpretation of this result is that their nervous systems possess oscillators which maintain information about the time of day and maintain it for at least two 24-hr cycles after the external cues have been removed.

This system may contain elements that are useful in understanding aspects of learning. This is particularly likely since Strumwasser (1963, 1965, 1967*a,b*) has shown a circadian rhythm in the impulse activity of a particular neuron of the parietovisceral ganglion in the same species. The neuron in question, variously designated as Br (Arvanitaki and Chalazonitis, 1956), Parabolic Burster (PB) 3 (Strumwasser, 1965), R15 (Frazier *et al.*, 1967),

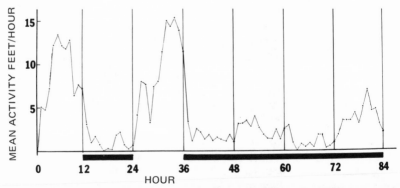

Fig. 26. Circadian locomotor rhythms in *Aplysia* (Kupfermann, 1968). Locomotion showed marked peaks during the periods of light, and these peaks recurred in less obvious form during two subsequent 24-hr cycles of darkness during the time of expected light.

and Oberon (Tauc, personal communication), all apparently the same neuron, exhibits a rhythm in its pattern of bursting having three recognizable components. First, over a period of minutes, intracellular records show bursts of from five to 30 impulses at 0.5–5/sec superimposed on depolarizing waves and with interburst intervals which range from a few seconds to a few minutes (Fig. 27). A second rhythm is manifest in the frequency and duration of these bursts. If a long-term continuous recording is made from the cell in a preparation taken from an animal that has been exposed to several days of alternating 12 hr of light and 12 hr of darkness, then an increase in the burst activity just described, or a long continuous burst, or both occur shortly after the projected dawn for at least two or three 24-hr cycles after removal of ganglion from the animal (Fig. 27). A third rhythm is expressed in the exact time of occurrence of the peak of activity, normally occurring near dawn. If the clock time of the peak in the activity is compared with the tidal cycles to which the animal had been exposed in its natural environment, a variation is apparent which parallels the tidal cycle with a slight lag.

There is no evidence as yet that the short-term bursting or the lunar tidal cycle in the expression of the bursting peaks is susceptible to plastic changes on the basis of stimulation temporarily applied either to the animal or to its nervous system. The circadian rhythm in the expression of the bursting peak, however, has elements which resemble rudimentary learning. For instance, the sharpness of the peaks in spike activity as well as their amplitude seems to depend on the number of conditioning cycles of light and dark given to the animal before the nervous system was isolated. The number of days over which this peak can be detected, a corollary of the above, depends also on the number of days of conditioning stimulation. Although there has not been any substantial progress in understanding the cellular or electrophysiological basis of the circadian rhythm in the single neuron, Strumwasser (1965, 1967a) has developed models for the generation of the microburst, i.e., the short-term bursting cycle, which may eventually be enlarged to include longer-term phenomena.

Neurons producing bursts resembling those of R15 in *Aplysis* have been found in *Helix* (Parmentier and Case, personal communication), in *Helisoma* (Kater and Kaneko, 1972), and in *Tritonia* (unpublished observation), but to date the circadian aspects of these bursting cycles have not been examined. In *Aplysia*, several other cells are known which produce bursts having a comparable time course to R15 (Wachtel, personal communication), but none of these have been studied in terms of their potential for longer-term variation or entrainment by external stimulation. It is not yet clear whether the rhythmic bursts play any behavioral role or even whether they can be considered a normal property of these neurons. The finding of Stinnakre and Tauc (1969) and Jahan-Parvar *et al.* (1969) that the bursts in *Aplysia* only appear *in vitro*,

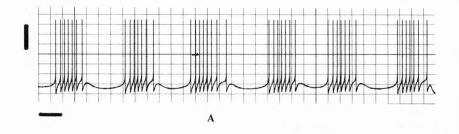

A

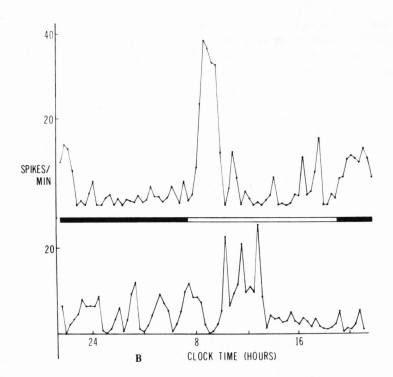

B   CLOCK TIME (HOURS)

Fig. 27. Cyclic bursting in neuron R15 of the parietovisceral ganglion of *Aplysia*. Upper L:
A commonly observed pattern of bursting in R15 recorded in daylight. Typically, although
not always, interspike intervals through each burst went from long to short to long again,
and a marked hyperpolarizing wave separated the bursts. Calibration was 30 mv and 10
sec. Lower: Average firing rate in R15 during two successive 24-hr cycles in isolated ganglion preparation (Strumwasser, 1965). Animal had earlier been subjected to nine successive 12-hr light – 12-hr dark cycles. Ganglion was then removed and neuron impaled.
Peaks in activity, made up of higher-frequency firing and bursts, and longer bursts,
occurred for at least two 24-hr cycles at or just after the time of expected dawn. Times of
projected light and dark are shown by light and dark bars.

and in particular when the connective nerve to the branchial ganglion is severed, raises some doubt as to the significance of the bursts for the whole animal.

C. Correlated Behavioral Learning and Neurophysiological Studies of Habituation and Sensitization

The essential element missing from the neural analogue studies of both habituation and classical conditioning in the work described earlier of Bruner and Tauc (1964, 1965a,b) and Kandel and Tauc (1965a,b) is that, despite the evident changes in neural activity resembling either habituation, sensitization, or classical conditioning, in no case was the specific behavioral or functional role of the neuron in question known. For this reason, the reports of Pinsker et al. (1970), Kupfermann et al. (1970), Castellucci et al. (1970), and Carew et al. (1972) have had a dramatic impact on the understanding of learning in gastropods. They shed new light on *Aplysia*'s behavioral capabilities, and perhaps more importantly they bolster confidence that the mechanisms proposed for hetero- and homosynaptic changes may be relevant to behavioral classical conditioning and habituation, respectively. These papers include analyses of both short- and long-term habituation of the withdrawal reflexes in *Aplysia* and the neuronal correlates of the short-term changes.

Pinsker et al. (1970) showed that whenever the mantle shelf or the base of the siphon was stimulated with a brief jet of water, the nearby gill was withdrawn. Furthermore, with repeated stimulation, at interstimulus intervals from one to a few minutes, the response amplitude declined to a fraction of its original control value over a period of ten to 20 stimuli. If stimulation was discontinued for an hour or two, the response level recovered to nearly that of the control. If a fully habituated animal was dishabituated by a strong and prolonged stimulus to the neck region between two applications of the jet of water to the siphon, the response level recovered to approximately two-thirds of the control level. It then rehabituated with continued stimulation. Recovery of the response to about 80 % of its original value took place in less than 20 min. The rate of habituation of the reflex was greater when the stimulus strength was less. The amplitudes of *spontaneous* gill withdrawals were not affected by habituation of the reflex, indicating that fatigue of the motor path or muscles themselves is not the likely cause. In addition, the habituation was greater with short rather than long interstimulus intervals, and the dishabituatory stimulus became less effective with repeated presentations. In all, this response showed at least six out of the widely quoted nine fundamental properties of habituation designated by Thompson and Spencer (1966).

An essential step in determining the neuronal mechanism of the gill reflex is the demonstration that habituation is taking place in the central nervous system and that the appropriate neurons can be identified and proven to be causing the observed changes. This is particularly true because the gill itself has a peripheral innervation and can habituate to directly applied stimulation in the absence of all central ganglion (Peretz, 1970). Earlier work by Kupfermann and Kandel (1969) and by Peretz (1969) has implicated several neurons in the abdominal ganglion in the control of gill movements. However, the specific neuronal correlates of both habituation and dishabituation of the gill withdrawal reflex were provided by Kupfermann *et al.* (1970). Four particular motor neurons, whose cell bodies are located in the left caudal quarter ganglion of the abdominal ganglion, had been shown to drive gill movements and to be sensitive to stimulation of the siphon or the mantle shelf. The critical point of this study was to establish whether or not the observed behavioral habituation could be attributed to changes observable in these motor neurons or whether, alternatively, the changes were taking place in sensory neurons preceding them or in the peripheral motor elements.

Recording from the motor neurons was accomplished in a semi-intact preparation. The abdominal ganglion was immobilized on a small platform placed beneath it. The possibility that changes were taking place either in the muscles or at the neuromuscular junctions was eliminated by directly stimulating the motor neurons and measuring the response of the gill under two sets of conditions. First, when they applied stimulation to the motor neuron at a rate comparable to the rates of sensory stimulation which produced complete habituation, Kupfermann *et al.* found that the motor response recorded in the gill did not decline appreciably. Second, when the response was fully habituated via the normal sensory stimulation, direct depolarization of the motor neuron produced action potentials which caused a normal gill retraction. They concluded that peripheral motor factors did not lie at the root of either habituation or dishabituation of the gill reflex.

The sensory side required a less direct test, because at that time the location of the sensory interneurons exciting the motor neurons was not known. However, three lines of evidence all pointed to a central focus for the observed behavioral habituation. First, if the afferent nerve trunk containing the inputs from the syphon and mantle shelf was directly stimulated electrically with a constant-amplitude shock, the gill withdrawal decremented in the usual way, thereby showing that the central nervous system could mediate habituation (even if under normal circumstances it did not). A second approach was to divide the afferent nerve into smaller bundles and record from the afferent units carrying the sensory nervous signals to look for a decline in their activity with repeated stimulation. Although the number of spikes seen varied widely from trial to trial, there was no systematic decline with as many as ten tactile stimuli presented every 1 or 2 min. Finally, these two indirect tests were corroborated using another technique. Kupfermann *et al.* first stimulated the siphon at the usual rates and strengths until the EPSP in the motor neuron declined to less than one-half of its control value. Then the sensory nerve carrying the stimulation to the nervous system was blocked by bathing it in a solution containing no ionic sodium or calcium while continuing the sensory stimulation to the peripheral receptors for 10–20 min. When the nerve was unblocked and the normal bathing solution returned to the nerve, the EPSP in the motor neuron reappeared with an amplitude that even exceeded the habituated response previous to the application of the block. Therefore, the sensory receptors were probably not adapting to the repeated stimuli.

The mechanism of the reduction in the amplitude of the EPSP was analyzed by Castellucci *et al.* (1970). In all its details, the phenomenon resembled that previously described for homosynaptic decrement by Bruner and Tauc (1965*b*). The very important element of progress was the successful reduction of the reflex to its monosynaptic elements including both sensory and motor elements of the pathway. The unitary EPSPs were depressed in amplitude with repeated stimulation of the appropriate patch of skin and showed a recovery resembling HSF after application of a dishabituating stimulus via another nerve. Apparently, the amount of transmitter being released by the sensory neuron at its synapse with the motor neuron is depressed by repeated stimulation through the same pathway.

Several other conclusions were drawn by the authors: (1) Dishabituation is probably the result of temporary facilitation mediated presynaptically, analogous to HSF. (2) This change in behavior probably does not require the formation of new synaptic contacts but rather is brought about by a change in the efficacy of synaptic contacts already present. (3) Inhibition of the motor neuron itself, as suggested by Holmgren and Frenk (1961), is not likely to be the cause of habituation of the gill withdrawal reflex. (4) Habituation and dishabituation are probably distinct phenomena, and dishabituation does not represent merely the decline of the habituated state.

This last conclusion was confirmed and enlarged upon by Carew *et al.* (1971), who showed, using essentially the same system but an additional sensory input pathway that could be habituated independently of the siphon, that dishabituation and sensitization are likely to be different manifestations of a common facilitory mechanism. In both cases, the amount of transmitter released would be increased at the presynaptic terminals, or alternatively the sensitivity of the postsynaptic receptors for the transmitter substance might be increased.

VI. SUMMARY AND CONCLUSIONS

When the behavioral and sensory capabilities of gastropods are taken fully into consideration and when the learning experiments already tried are reviewed, it is obvious that only a very limited range out of the total very broad spectrum of possibilities has been attempted. Furthermore, in a great number of cases, the behaviors least likely to be plastic on an *a priori* basis are the ones most intensively studied. For instance, although it is clear that the chemical senses of gastropods are exceedingly keen and probably capable of subtle discriminations, relatively few learning experiments have taken advantage of this fact. On the other hand, the visual capabilities of gastropods are almost certainly very limited, but nevertheless numerous attempts have been made to use visual stimulation as a cue in learning experiments. The greatest majority of all learning experiments reported to date have been attempts to modify simple withdrawal reflexes or escape patterns, while very few have endeavored to change any of the numerous action patterns relating to homing, boring, brooding of egg capsules, courtship, copulatory behavior, or the various aspects of food location and feeding activity.

Taken together, the evidence for learning in gastropods, with the exception of habituation and sensitization, is not impressive. On the other hand, the behavioral and neurophysiological experiments with habituation and sensitization have been numerous and are, on the whole, convincing. This is particularly true in experiments performed since Humphrey broadened the base for experimental testing of habituation in gastropods in 1930. Even before that, however, it is clear that Pieron understood the importance of separating peripheral phenomena from central ones. Likewise, sensitization has been unambiguously demonstrated in several attempts at classical conditioning, and in fact may eventually prove to underlie what has been described as classical conditioning in certain other experiments.

In all, there have been not more than three or four studies of classical conditioning with positive interpretations, and of these not all were done blind to eliminate experimenter bias and not all had adequate controls for sensitization. However, an equally limited number of experiments have been reported which had negative results for classical conditioning. In each of these cases, a positive result might have been reported had not the sensitization controls been carried out. Furthermore, one cannot know how many unsuccessful attempts have been made to classically condition animals but have not been reported in the literature. Inquiries to many of the authors whose work is described here did not suggest that this number is large, but it is quite obvious that a positive result is often more exciting and reinforcing to the experimenter than a negative one and may more often lead to published work. Yet a wholly negative conclusion in regard to classical conditioning in gastropods is certainly premature, since virtually all of the unsuccessful experiments have been attempts to classically condition escape response behaviors for which there may be great adaptive value in stereotypy and "hard wiring." The reports of successful experiments have more often involved feeding.

It is interesting and may be significant that the aggregates of behavioral evidence for habituation, for sensitization, and for classical conditioning are approximately as good as the electrophysiological data for an analogue at the neuronal level in each case. While both habituation and sensitization are well established behaviorally and their neurophysiological bases have been studied, classical conditioning in behavior is as yet not fully supported by experimental evidence, and analysis of the specific form of heterosynaptic facilitation is similarly incomplete.

Passive avoidance learning and operant conditioning reported by Lickey (1968) and Lee (1969, 1970), respectively, are both convincing, but it is difficult to point to any other studies, particularly from the literature on maze learning, which support a well-developed associative learning capability is gastropods. The failure of the aquatic forms to learn mazes, even as simple as T- and Y-mazes, is all the more surprising in view of the apparent ability of certain marine forms to return to their home bases or to their egg capsules, a task which would seem to require a similar sort of learning on a more complex level. A factor which may underlie this apparent discrepancy arises out of differences observed in laboratory *vs.* field behavior. The laboratory behavior of snails such as *Fusitriton* is in many respects grossly distorted by comparison with their responses and capabilities in the field. The maze-learning tasks have, with the exception of one of the attempts of Fischel (1931), all been carried out in highly artificial laboratory surroundings, and even in the case just mentioned the experimental apparatus was only an approximation to the normal environment. The suggestion follows then that preliminary fieldwork and variations in laboratory conditions to determine the normal capabilities of animals would almost certainly prove to be worthwhile, particularly in regard to classical and operant conditioning attempts.

One may take the view that it is not instructive to evaluate whether, out of the whole spectrum of gastropod behavioral possibilities, there is some task or some discrimination or some association that some gastropods can learn successfully on a few occasions. As has been pointed out by Wells (1965), "Tested under appropriate conditions, most animals can be shown to learn something sometime." The question is more realistically put in terms of the overall capability of gastropods to take advantage of learning as a normal and adaptive process in dealing with their environment. The only kinds of plasticity which gastropods can unequivocally be said to use within the context of this sort of a requirement are habituation and sensitization, both of which are well established. The associative forms of learning which require the correlation of different stimuli occurring separately in time are not well established experimentally and may not occur widely in the gastropod class.

REFERENCES

Abbott, D. P., 1962, Observations on the gastropod *Terebellum terebellum* (Linnaeus) with particular reference to the behavior of the eyes, during burrowing, *Veliger, 5*, 1–3.

Abraham, F. D., and Willows, A. O. D., 1971, Plasticity of a fixed action pattern of behavior in the sea slug *Tritonia diomdia, Comm. Behav. Biol. A6*, 271–280.

Agersborg, H. P. K., 1922, Notes on the locomotion of the nudibranchiate mollusc *Dendronotus giganteus* O'Donoghue, *Biol. Bull., 42*, 257–265.

Agersborg. H. P. K., 1923, The morphology of *Melibe leonina, Quart. J. Microscop. Sci., 67*, 507–592.

Allen, E. J., and Nelson, E. W., 1911, On the artificial culture of marine organisms, *J. Marine Biol. Asso. U.K., 8*, 421–474.

Amoroso, E. C., Baxter, M. I., Chiquoine, A. D., and Nisbet, R. H., 1964, The fine structure of neurons and other elements in the nervous system of the giant African land snail *Archachatina marginata, Proc. Roy. Soc. Ser. B, 160*, 167–180.

Ankel, W. E., 1936, Prosobranchia, *in* "Die Tierwelt der Nord- und Ostsee," Vol. 9, Akademische Verlagsgesellschaft, Leipzig.

Ansell, A. D., 1969, Defensive adaptations to predation in the Mollusca, *Symp. Marine Biol. Ass. India*, pp. 487–512.

Arey, L. B., and Crozier, W. J., 1919, The nervous organization of a nudibranch, *Proc. Natl. Acad. Sci., 5*, 498–500.

Arey, L. B., and Crozier, W. J., 1921, On the natural history of *Onchidium, J. Exptl. Zool., 32*, 443–502.

Arnold, D. C., 1957, The response of the limpet *Patella vulgata* L., to waters of different salinities, *J. Marine Biol. Ass. U.K., 36*, 121–128.

Arvanitaki, A., and Chalazonitis, N., 1956, Activations du soma géant d'*Aplysia* par voie antidrome (derivation endocytaire), *Arch. Sci. Physiol., 10*, 95–128.

Arvanitaki, A., and Chalazonitis, N., 1960, Photopotentials d'excitation et d'inhibition de differents somata identifiables (*Aplysia*). Activations monochromatiques, *Bull. Inst. Oceanogr. (Monaco), 57*, 1–83.

Arvanitaki, A., and Chalazonitis, N., 1961, Excitatory and inhibitory processes initiated by light and infrared radiations in single identifiable nerve cells (giant ganglion cells of *Aplysia*), *in* "Nervous Inhibition" (E. Florey, ed.), pp. 194–231, Pergamon Press, Oxford.

Bailey, D. S., and Laverack, M. S., 1963, Central responses to chemical stimulation of a gastropod osphradium, *Nature (Lond.), 200*, 1122–1123.

Bailey, D. S., and Laverack, M. S., 1966, Aspects of the neurophysiology of *Buccinum undatum* L. (Gastropoda). I. Central responses to stimulation of the osphradium. *J. Exptl. Biol., 44*, 131–148.

Barraud, E. M., 1957, The copulatory behaviour of the freshwater snail (*Limnaea stagnalis* L.), *Brit. J. Anim. Behav., 5*, 55–59.

Barth, J., 1964, Intracellular recordings from photosensory neurons in the eyes of a nudibranch mollusc (*Hermissenda crassicornis*), *Comp. Biochem. Physiol., 11*, 311–315.

Berg, C. J., 1971, Ontogeny of the behavior of a strombid gastropod, *Am. Zoologist, 11,* 640.

Bethe, A., 1903, "Allegemeine Anatomie und Physiologie des Nervensystems," Thieme, Leipzig.

Born, E., 1910, Beiträge zur feineren Anatomie der *Phyllirhae bucephala, Z. Wiss. Zool., 97,* 105–192.

Bovbjerg, R. V., 1965, Feeding and dispersal in the snail *Stagnicola reflexa* (*Basommatophora: Lymnaeidae*), *Malacologia, 2,* 199–207.

Bovbjerg, R. V., 1968, Responses to food in lymnaid snails, *Physiol. Zool., 41,* 412–423.

Brown, A. C., 1961, Physiological–ecological studies on two sandy-beach Gastropoda from South Africa; *Bullia digitalis* Meuschen and *Bullia laevissima* (Ginelin), *Z. Morphol. Okol. Tiere, 49,* 629–657.

Brown, A. C., and Noble, R. G., 1960, Function of the osphradium in *Bullia* (Gastropoda), *Nature* (*Lond.*), *188,* 1045.

Bruner, J., and Tauc, L., 1964, Les modifications de l'activité synaptique au cours d'habituation chez l'*Aplysie, J. Physiol.* (*Paris*), *56,* 306.

Bruner, J., and Tauc, L., 1965*a,* La plasticité synaptique impliquée dans le processus d'habituation chez l'*Aplysie, J. Physiol.* (*Paris*), *57,* 230.

Bruner, J., and Tauc, L., 1965*b,* Long-lasting phenomena in the molluscan nervous system, *Symp. Soc. Exptl. Biol., 20,* 457–475.

Bullock, T. H., 1953, Predator recognition and escape responses of some intertidal gastropods in the presence of starfish, *Behaviour, 4–6,* 130–146.

Burdon-Jones, C., and Charles, G. H., 1958, Light reactions of littoral gastropods, *Nature* (*Lond.*)., *181,* 129–131.

Buytendijk, F. J. J., 1921, Une formation d'habitude simple chez le limacon d'eau douce (*Limnaea*), *Arch. Neerl. Physiol., 5,* 458–466.

Carew, T. J., Castellucci, V. F., and Kandel, E. R., 1971, An analysis of dishabituation and sensitization of the gill-withdrawal reflex in *Aplysia, Internat. J. Neurosci., 2,* 79–98.

Carew, T. J., Pinsker, H. M., and Kandel, E. R., 1972, Long-term habituation of a defensive withdrawal reflex in *Aplysia, Science, 175,* 451–454.

Castellucci, V., Pinsker, H., Kupfermann, I., and Kandel, E. R., 1970, Neuronal mechanisms of habituation and dishabituation of the gill-withdrawal reflex in *Aplysia, Science, 167,* 1745–1748.

Charles, G. H., 1966, *in* "Physiology of Mollusca" (K. M. Wilbur and C. M. Yonge, eds.), Vol. 2, pp. 455–521, Academic Press, New York.

Chase, L., 1953, The aerial mating of the great slug, *Discovery, 13,* 356–359.

Clark, R. B., 1964, "Dynamics in Metazoan Evolution," pp. 53–64, Clarendon Press, Oxford.

Coggeshall, R. E., Yaksta, B. A., and Swartz, F. J., 1970, A cytophotometric analysis of the DNA in the nucleus of the giant cell R-2, in *Aplysia, Chromosoma, 32,* 205–212.

Colton, H. S., 1918, Fertilization in the air-breathing pond snails, *Biol. Bull., 35,* 48–49.

Cook, A., 1971, Habituation in a freshwater snail (*Limnaea stagnalis*), *Anim. Behav., 19,* 463–474.

Cook, A., Bamford, O. S., Freeman, J. D. B., and Teideman, D. J., 1969, A study of the homing habit of the limpet, *Anim. Behav., 17,* 330–339.

Cook, S. B., 1969, Experiments on homing in the limpet *Siphonaria normalis, Anim. Behav. 17,* 679–682.

Copeland, M., 1918, The olfactory reactions and organs of the marine *Alectrion obseolata* (Say) and *Busycon canaliculatum Lin., J. Exptl. Zool., 25,* 177–227.

Crabb, E. D., 1927, The fertilization process in the snail *Lymnaea stagnalis appressa* Say, *Biol. Bull., 53,* 67–108.

Crisp, M., 1969, Studies on the behaviour of *Nassarius obsoletus, Biol. Bull., 136*, 355.

Crozier, W. J., and Arey, L. B., 1919, Sensory reactions of *Chromodoris zebra, J. Exptl. Zool., 29*, 261–310.

Dakin, W. J., 1910, The visceral ganglion of *Pecten*, with some notes on the physiology of the nervous system and an inquiry into the innervation of the osphradium in the Lamellibranchiata, *Mitt. Zool. Staz. Neapel, 20*, 1–40.

D'Asaro, C., 1965, Organogenesis, development and metamorphosis in the Queen conch, *Strombus gigas*, with notes on breeding habits, *Bull. Marine Sci., 15*, 359–416.

Davis, J. R. A., 1895, The habits of limpets, *Nature* (*Lond.*), *51*, 511–512.

Dawson, J., 1911, The biology of *Physa, Behav. Monogr., 1*, No. 4.

Dennis, M. J., 1967, Interactions between the five receptor cells of a simple eye (in *Hermissenda crassicornis*), *in* "Conference on Invertebrate Nervous Systems: Their Significance for Mammalian Neurophysiology" (C. A. G. Wiersma, ed.), pp. 259–262, University of Chicago Press, Chicago.

Dorsett, D. A., Willows, A. O. D., and Hoyle, G., 1969, Centrally generated nerve impulse sequences determining swimming behavior in *Tritonia, Nature* (*Lond.*), *224*, 711–712.

Dorsett, D. A., Willows, A. O. D., and Hoyle, G., 1973, The neuronal basis of behavior in *Tritonia*. IV. The central origin of a fixed action pattern demonstrated in the isolated brain, *J. Neurobiol.*, in press.

Edelstam, C., and Palmer, C., 1950, Homing behaviour in gastropods, *Oikos, 2*, 259–270.

Edmunds, M., 1968, On the swimming and defensive response of *Hexabranchus marginatus* (Mollusca, Nudibranchia), *J. Linn. Soc. Zool., 47*, 425–429.

Farmer, W. M., 1970, Swimming gastropods (Opisthobranchia and Prosobranchia), *Veliger, 13*, 73–80.

Feder, H. M., and Christensen, A. M., 1966, Aspects of asteroid biology, *in* "Physiology of Echinodermata" (Richard A. Boolootian, ed.), pp. 87–127, Interscience (Wiley), New York.

Fischel, W., 1931, Dressurversuche an Schnecken, *Z. Vergl. Physiol., 15*, 50–70.

Fischer, H., 1898, Quelques remarques sur les moeurs des Patelles, *J. Conchylol., 46*, 314–318.

Fisher-Piette, E., 1935, Histoire d'une mouliére. Observations sur une phase de déséquilibre faunique, *Bull. Biol. France Belg., 69*, 165–177.

Frankel, G., 1927, Geotaxis und Phototaxis von *Littorina, Z. Vergl. Physiol., 5*, 585–597.

Frazier, W. T., Waziri, R., and Kandel, E. R., 1965, Alterations in the frequency of spontaneous activity in *Aplysia* neurons with contingent and non-contingent nerve stimulation, *Fed. Proc., 24*, 522.

Frazier, W. T., Kandel, E. R., Kupfermann, I., Waziri, R., and Coggeshall, R. E., 1967, Morphological and functional properties of identified neurons in the abdominal ganglion of *Aplysia californica, J. Neurophysiol., 30*, 1288–1351.

Fretter, V., and Graham, A., 1962, "British Prosobranch Mollusks," 744 pp., Ray Society, London.

Fretter, V., and Graham, A., 1964, *in* "Physiology of Mollusca" (K. M. Wilbur and C. M. Yonge, eds.), Vol. 1, pp. 127–164, Academic Press, New York.

Fretter, V., and Montgomery, M. C., 1968, The treatment of food by prosobranch veligers, *J. Marine Biol. Ass. U.K., 48*, 499–520.

Frings, H., and Frings, C., 1965, Chemosensory bases of food-finding and feeding in *Aplysia juliana* (Mollusca, Opisthobranchia), *Biol. Bull., 128*, 211–217.

Garth, T. R., 1924, The learning curve for a snail, *Science, 59*, 440.

Garth, T. R., and Mitchell, M. P., 1926, The learning curve of a land snail, *J. Comp. Physiol. Psychol., 6*, 103–113.

Giller, E., and Schwartz, J. H., 1968, Choline acetyltransferase: Regional distribution in the abdominal ganglion of *Aplysia, Science, 161*, 908.

Gorman, A. L. F., and Mirolli, M., 1969, The input–output organization of a pair of giant neurons in the mollusc, *Anisodoris nobilis* (MacFarland), *J. Exptl. Biol., 51*, 615–634.

Hewatt, W. G., 1940, Observations on the homing limpet, *Acmaea scabra* Gould, *Am. Midl. Naturalist, 24*, 205–208.

Holmgren, B., and Frenk, S., 1961, Inhibitory phenomena and "habituation" at the neuronal level, *Nature (Lond.), 192*, 1294–1295.

Howells, H. H., 1942, The structure and function of the alimentary canal of *Aplysia punctata, Quart. J. Microscop. Sci., 83*, 357–396.

Hoyle, G., and Willows, A. O. D., 1973, The neuronal basis of behavior in *Tritonia*. II. The relationship of muscular contraction to nerve impulse pattern, *J. Neurobiol.,* in press.

Hughes, G. M., 1967, The left and right giant neurons (LGC and RGC) of *Aplysia, in* "Neurobiology of Invertebrates" (J. Salanki, ed.), pp. 423–441, Plenum Press, New York.

Hughes, G. M., and Tauc, L., 1963, An electrophysiological study of the anatomical relations of two giant nerve cells in *Aplysia depilans, J. Exptl. Biol., 40*, 469–486.

Humphrey, G., 1930, Le Chatelier's rule, and the problem of habituation and dehabituation in *Helix albolabris, Psychol. Forsch., 13*, 113–127.

Hurst, A., 1968, The feeding mechanism and behaviour of the opisthobranch, *Melibe leonina, Symp. Zool. Soc. Lond., 22*, 151–166.

Hutchinson, G. E., 1930, Two biological aspects of psycho-analytic theory, *Internat. J. Psychoanal., 1*, 83–86.

Jahan-Parvar, B., 1970, Conditioned response in *Aplysia californica, Am. Zoologist, 10*, 287.

Jahan-Parvar, B., 1971, Neuronal analogs of classical conditioning in *Aplysia californica, Proc. Internat. Union Physiol. Sci., 9* (abst.).

Jahan-Parvar, B., and von Baumgarten, R. J., 1967, Untersuchungen zur Spezifitätsfrage der heterosynaptischen Facilitation bei *Aplysia californica, Pflugers Arch. Ges. Physiol., 295*, 347–360.

Jahan-Parvar, B., Smith, M., and von Baumgarten, R., 1969, Activation of neurosecretory cells in *Aplysia* by osphradial stimulation, *Am. J. Physiol., 216*, 1246–1257.

Jones, H. D., and Trueman, E. R., 1970, Locomotion of the limpet, *Patella vulgata, J. Exptl. Biol., 52*, 201–216.

Kandel, E. R., and Spencer, W. A., 1968, Cellular neurophysiological approaches in the study of learning, *Physiol. Rev., 48*, 65–134.

Kandel, E. R., and Tauc, L., 1965*a*, Heterosynaptic facilitation in neurons of the abdominal ganglion of *Aplysia depilans, J. Physiol. (Lond.), 181*, 1–27.

Kandel, E. R., and Tauc, L., 1965*b*, Mechanisms of heterosynaptic facilitation in the giant cell of the abdominal ganglion of *Aplysia depilans, J. Physiol. (Lond.), 181*, 28–47.

Karnaukhov, V. N., 1971, Carotenoids in oxidative metabolism of molluskan neurons, *Exptl. Cell Res., 64*, 301–306.

Kater, S. B., and Kaneko, 1972, An endogenously bursting neuron in the gastropod mollusc, *Helisoma trivolvis*. Characterization of activity, *In Vivo, J. Comp. Physiol. 79*, 1–14.

Kater, S. B., Heyer, C., and Hegmann, J. P., 1971, Neuromuscular transmission in the gastropod mollusc *Helisoma trivolvis:* Identification of motorneurons, *Z. Vergl. Physiol., 74*, 127–139.

Kennedy, D., 1960, Neural photoreceptors in a lamellibranch mollusc, *J. Gen. Physiol., 44*, 227–299.

Kohn, A. J., 1961, Chemoreception in gastropod molluscs, *Am. Zoologist, 1*, 291–308.

Kristan, W. B., 1971, Plasticity of firing patterns in neurons of *Aplysia* pleural ganglion, *J. Neurophysiol., 34*, 321–336.

Kristan, W. B., and Gerstein, G. L., 1970, Plasticity of synchronous activity in a small neural net, *Science, 169*, 1336–1339.

Kuhlmann, D., 1969, Bestimmung des DNS-Gehaltes in Zellkernen des Nervengewebes von *Helix pomatia* L. und *Planorbarius corneus* L. (Stylommatophora und Basommatophora, Gastropoda), *Experientia, 25(8)*, 848–849.

Kupfermann, I., 1968, A circadian locomotory rhythm in *Aplysia californica, Physiol. Behav., 3*, 179–181.

Kupfermann, I., and Kandel, E. R., 1969, Neuronal controls of a behavioral response mediated by the abdominal ganglion of *Aplysia, Science, 164*, 847–850.

Kupfermann, I., and Pinsker, H., 1969, Plasticity in *Aplysia* neurons and some simple neuronal models of learning, *in* "Reinforcement" (J. Tapp, ed.), pp. 356–386, Academic Press, New York.

Kupfermann, I., and Pinsker, H., 1970, Cellular models of learning and cellular mechanisms of plasticity in *Aplysia, in* "Biology of Memory," pp. 163–174, Academic Press, New York.

Kupfermann, I., Castellucci, V., Pinsker, H., and Kandel, E. R., 1970, Neuronal correlates of habituation and dishabituation of the gill withdrawal reflex in *Aplysia, Science, 167*, 1743–1745.

Kupfermann, I., Pinsker, H., Castellucci, V., and Kandel, E. R., 1971, Central and peripheral control of gill movements in *Aplysia, Science, 174*, 1252–1255.

Lasek, R. J., and Dower, W. J., 1971, *Aplysia californica:* Analysis of nuclear DNA in individual nuclei of giant neurons, *Science, 172*, 278–280.

Laverack, M. S., and Bailey, D. F., 1963, Movement receptors in *Buccinum undatum, Comp. Biochem. Physiol., 8*, 289–298.

Laxton, J. H., 1971, Feeding in some Australasian Cymatiidae (Gastropoda: Prosobranchia), *Zool. J. Linn. Soc., 50*, 1–9.

Lebour, M. V., 1937, The eggs and larvae of the British prosobranchs with special reference to those living in the plankton, *J. Marine Biol. Ass. U.K., 22*, 105–166.

Lee, R. M., 1969, Effects of contingent water level variation, *Comm. Behav. Biol., 3*, 157–164.

Lee, R. M., 1970, *Aplysia* behavior: Operant-response differentiation, *Proc. Am. Psychol. Ass.,* Div. *6*, 249–250.

Lemche, H., 1957, A new living deep-sea mollusc of the cambro-Devonian class, Monoplacophora, *Nature (Lond.), 179*, 413–416.

Lickey, M. E., 1968, Learned behavior in *Aplysia vaccaria, J. Comp. Physiol. Psychol., 66*, 712–718.

Lickey, M. E., and Berry, R. W., 1966, Learned behavioral discrimination of food objects by *Aplysia californica, Physiologist, 9*, 230.

Lissman, H. W., 1945a, The mechanism of locomotion in gastropod molluscs. I. Kinematics, *J. Exptl. Biol., 21*, 58–69.

Lissman, H. W., 1945b, The mechanism of locomotion in gastropod molluscs. II. Kinetics, *J. Exptl. Biol., 22*, 37–50.

Lukowiak, K., and Jacklet, J. W., 1972, Habituation and dishabituation: Interactions between peripheral and central nervous systems in *Aplysia, Science, 178*, 1306–1308.

MacGinitie, G. E., 1934, The egg-laying activities of the sea-hare, *Tethys californicus* (Cooper), *Biol. Bull., 67*, 300–103.

Margolin, A. S., 1964, A running response of *Acmaea* to seastars, *Ecology, 45*, 191–193.

McCaman, R. E., and Dewhurst, S. A., 1971, Metabolism of putative transmitters in individual neurons of *Aplysia californica, J. Neurochem., 18*, 1329–1336.

Medioni, J., 1959, Étude de la sensibilité visuelle de *Limnaea stagnalis* par la méthode de la réaction skioptique, *Compt. Rend. Soc. Biol. Paris, 152*, 840.

Michelson, E. H., 1960, Chemoreception in the snail *Australorbis glabratus, Am. J. Trop. Med. Hyg., 9*, 480–487.

Miller, S. E., 1969, Locomotion, foot form and tenacity in the Muricacea (Gastropoda), Masters thesis, University of Washington.

Morton, J. E., 1954, The biology of *Limacina retroversa, J. Marine Biol. Ass. U. K., 33,* 297–312.

Morton, J. E., 1964, Locomotion, *in* "Physiology of Mollusca" (K. M. Wilbur and C. M. Yonge, eds.), Vol. 1, pp. 383–423, Academic Press, New York.

Morton, J. E., 1967, "Molluscs," 244 pp., Hutchinson University Library, London.

Morton, J. E., and Holme, N. A., 1955, The occurrence at Plymouth of the opisthobranch *Akera bullata* with notes on its habits and relationships, *J. Marine Biol. Ass. U.K., 34,* 101–112.

Newell, G. E., 1965, The eye of *Littorina littorea, Proc. Zool. Soc. Lond., 144,* 75–86.

Newell, P. F., 1966, The nocturnal behaviour of slugs, *Med. Biol. Ill., 16,* 146–159.

Nicaise, G., 1967, Neuro-interstitial junction in a gastropod, *Glossodoris, Nature (Lond.),* *216,* 1222–1223.

Nicaise, G., Pavans de Ceccatty, M., and Baleydier, C., 1968, Ultrastructures des connexions entre cellules nerveuses, musculaires et glio-interstitielles chez *Glossodoris, Z. Zellforsch., 88,* 470–486.

Nisbet, R. H., 1961, Some aspects of the neurophysiology of *Archachatina (Calachatina) marginata* (Swainson), *Proc. Roy. Soc. B., 154,* 309–331.

Owen, G., 1966 *in* "Physiology of Mollusca" (K. M. Wilbur and C. M. Yonge, eds.), pp. 1–51, Academic Press, New York

Paine, R. T., 1963, Food recognition and predation on opisthobranchs by *Navanax inermis* (Gastropoda: Opisthobranchia), *Veliger, 6,* 1–9.

Parker, G. H., 1917, The pedal locomotion of the sea-hare *Aplysia californica, J. Exptl. Zool., 24,* 139–145.

Parker, G. H., 1922, The leaping of the stromb, *Strombus gigas, J. Exptl. Zool., 24,* 139–145.

Peretz, B., 1969, Central neuron initiation of periodic gill movements, *Science, 166,* 1167–1172.

Peretz, B., 1970, Habituation and dishabituation in the absence of a central nervous system, *Science, 9,* 379–381.

Pieron, H., 1910, L'adaptation aux obscurations répétées comme phénomène de mémoire chez les animaux inférieurs. La loi de l'oubli chez la *Limnée, Arch. Psychol. Geneve, 9,* 39–50.

Pieron, H., 1911, Sur la détermination de la periode d'établissement dans les acquisitions mnémoniques, *Compt. Rend. Acad. Sci. Paris, 152,* 1410–1413.

Pieron, H., 1913, Recherches experimentales sur les phénomènes de mémorie, *Ann. Psychol., 19,* 91–193.

Pilkington, M. C., and Fretter, V., 1970, Some factors affecting the growth of prosobranch veligers, *Helgolander Wiss. Meeresunters., 20,* 576–593.

Pinsker, H., and Kandel, E. R., 1967, Contingent modification of an endogenous bursting rhythm by monosynaptic inhibition, *Physiologist, 10,* 279.

Pinsker, H., Kupfermann, I., Castellucci, V., and Kandel E. R., 1970, Habituation and dishabituation of the gill-withdrawal reflex in *Aplysia, Science, 167,* 1740–1742.

Preston. R. J., and Lee, R. M., 1972, Feeding behavior in *Aplysia californica:* Role of chemical and tactile stimuli, *J. Comp. Physiol. Psychol.,* in press.

Retzius, G., 1892, Das sensible Nervensystem der Mollusken, *Biol. Untersuch. (N.F.), 4,* 11–18.

Roach, D. K., 1967, Rhythmic muscular activity in the alimentary tract of *Arion ater* (L.) (Gastropoda: Pulmonata), *Comp. Biochem. Physiol., 24,* 865–878.

Scheltema, R. S., 1962, Pelagie larvae of New England intertidal gastropods. I. *Nassarius obsoletus* Say and *Nassarius videx* Say, *Trans. Am. Microscop. Soc., 81,* 1–11.

Schmidt-Nielsen, K., Taylor, C. R., and Shkolnik, A., 1971, Desert snails: Problems of heat, water and food, *J. Exptl. Biol., 55*, 385–398.

Smith, S. T., and Carefoot, T. H., 1967, Induced maturation of gonads in *Aplysia punctata* Cuvier, *Nature, 215*, 652–653

Snyder, N. F. R., and Snyder, H. A., 1971, Pheromone-mediated behaviour of *Fasciolaria tulipa, Anim. Behav., 19*, 257–268.

Sokolov, V. A., 1959, Conditioned reflex in *Physa acuta, Vestnik Leningrad Univ. Ser. Biol., 2(9)*, 82–86.

Stinnakre, J., and Tauc, L., 1966, Effects de l'activation de l'osphradium sur les neurones du système nerveux central de l'*Aplysie, J. Physiol. (Paris), 58*, 266–267.

Stinnakre, J., and Tauc, L., 1969, Central neuronal response to the activation of osmoreceptors in the ophradium of *Aplysia, J. Exptl. Biol., 51*, 347–361.

Strumwasser, F., 1963, A circadian rhythm of activity and its endogenous origin in a neuron, *Fed. Proc., 22*, 220.

Strumwasser, F., 1965, The demonstration and manipulation of a circadian rhythm in a single neuron, *in* "Circadian Clocks" (J. Aschoff, eds.), pp. 442–462, North-Holland, Amsterdam.

Strumwasser, F., 1967a, Membrane and intracellular mechanisms governing endogenous activity in neurons, *in* "Physiological and Biochemical Aspects of Nervous Integration" (F. D. Carlson, ed.), pp. 329–341, Prentice-Hall, Englewood Cliffs, N. J.

Strumwasser, F., 1967b, Neurophysiological aspects of rhythms, *in* "The Neurosciences" (G. C. Quarton, T. Melnechuk, and F. O. Schmitt, eds.), pp. 516–528, Rockefeller Univ. Press, N.Y.

Szal, R., 1971, New sense organ of primitive gastropods, *Nature (Lond.), 229*, 490–492.

Tardy, J., 1970, Contribution à l'étude des metamorphoses chez les nudibranches, *Ann. Sci. Nat. Zool., 12*, 299–370.

Tauc, L., and Epstein, R., 1967, Heterosynaptic facilitation as a distinct mechanism in *Aplysia, Nature (Lond.), 214*, 724–725.

Thompson, E. L., 1917, An analysis of the learning process in the snail, *Physa gyrina* Say, *Behav. Monogr., 3,*

Thompson. T. E., 1958, The natural history, embryology, larval biology, and post larval development of *Adalaria proxima* (Alder and Hancock) (Gastropoda Opisthobranchia), *Phil. Trans. Roy. Soc. Ser. B, 242*, 1–58.

Thompson, T. E., 1962, Studies on the ontogeny of *Tritonia hombergi* Cuvier (Gastropoda Opisthobranchia), *Phil. Trans. Roy. Soc. Ser. B, 245*, 171–218.

Thompson, T. E., 1964, Grazing and the life cycles of British nudibranchs, *in* "Fourth British Ecological Society Symposium (D. J. Crisp, ed.), pp. 274–297, Blackwell, Oxford.

Thompson, R. F., and Spencer, W. A., 1966, Habituation: A model phenomenon for behavior, *Psychol. Rev., 173*, 1643.

Thorpe, W. H., 1963, "Learning and Instinct in Animals," 558 pp., Methuen, London.

Thorson, G., 1946, Reproduction and larval development of Danish marine bottom invertebrates, *Medd. Komm. Havundersg. Kbh. (Ser. Plankton), 4*, 1–523.

Turner, R. D., 1959, Notes on the feeding of *Melongena corona, Nautilus, 73*, 11–13.

Veratti, E., 1900, Ricerche sul sistema nervoso dei *Limax, Mem. Ist. Lombardo, 18*, 167–179.

Verlaine, L., 1936, L'instinct et l'intelligence chez les mollusques. Les gastéropodes perceurs de coquilles, *Mem. Mus. Hist. Nat. Belg., 3*, 387–394.

von Baumgarten, R., and Hukuhara, T., 1969, The role of the interstimulus interval in heterosynaptic facilitation in *Aplysia californica, Brain Res., 16*, 369–381.

von Baumgarten, R., and Jahan-Parvar, B., 1967a, Beitrag zum Problem der heterosynaptischen Facilitation in *Aplysia californica, Pflügers Arch. Ges. Physiol., 295*, 328–346.

von Baumgarten, R., and Jahan-Parvar, B., 1967b, Time course of repetitive heterosynaptic facilitation in *Aplysia californica, Brain Res., 4,* 295–297.

von Buddenbrock, W., 1919, Analyse der Lichtreaktionen der Helicidien, *Zool. Jahrb., Abt. Allgem. Zool. Physiol. Tiere, 37,* 315–360.

Walne, P. R., 1964, *in* "Physiology of Mollusca" (K. M. Wilbur and C. M. Yonge, eds.), Vol. 1, pp. 197–210, Academic Press, New York.

Webb, M., Moodley, L. G., and Thandar, A. S., 1969, The copulatory mechanism in *Onchidium peronii, S. Afr. J. Sci., 65,* 107–112.

Wells, M. J., 1965, Learning by marine invertebrates, *Advan. Marine Biol.. 3,* 1–62.

Wells, M. J., and Wells, J., 1971, Conditioning and sensitization in snails, *Anim. Behav., 19,* 305–312.

Willows, A. O. D., 1967, Behavioral acts elicited by stimulation of single, identifiable brain cells, *Science, 157,* 570–574.

Willows. A. O. D., 1969, Neuronal network triggering a fixed action pattern, *Science, 166,* 1549–1551.

Willows, A. O. D., Dorsett, D. A., and Hoyle, G., 1973a, The neural basis of behavior in *Tritonia.* I. Functional organization of the central nervous system, *J. Neurobiol.,* in press.

Willows, A. O. D., Dorsett, D. A., and Hoyle, G., 1973b, The neural basis of behavior in *Tritonia.* III. Neuronal mechanism of a fixed action pattern, *J. Neurobiol.,* in press.

Wolff, H. G., 1969, Einige Ergebnisse zur Ultrastruktur der Statocysten von *Limax maximus, Limax flavus* und *Arion empiricorum* (Pulmonata), *Z. Zellforsch., 100,* 251–270.

Wolff, H. G., 1970, Statocystenfunktion bei einigen Landpulmonaten (Gastropoda), *Z. Vergl. Physiol., 69,* 326–366.

Wolper, C., 1950, Das Osphradium der *Paludina vivipara, Z. Vergl. Physiol., 32,* 272–286.

Wurtz, R. H., Castellucci, V. F., and Nusrala, J. M., 1967, Synaptic plasticity: The effect of the action potential in the post-synaptic neuron, *Exptl. Neurol., 18,* 350–368.

Yerkes, R. M., 1912, The intelligence of the earthworm, *J. Anim. Behav., 2,* 332.

Yom-Tov Y., 1970, The effect of predation on population densities of some desert snails, *Ecology, 51,* 907–911

Yonge, C. M., 1937, The biology of *Aporrhais pas-pelicani* (L.) and *A. serresiana* (Mich.), *J. Marine Biol. Ass. U.K., 21,* 687–703.

Yonge, C. M., 1938, Evolution of ciliary feeding in the Prosobranchia with an account of feeding in *Copulus ungaricus, J. Marine Biol. Ass. U.K., 22,* 453–468.

Yonge, C. M., 1947, The pallial organs in the aspidobranch Gastropoda and their evolution throughout the Mollusca, *Phil. Trans. Roy. Soc. Ser. B, 232,* 443–518.

Young, J. Z., 1962, The retina of cephalopods and its degeneration after optic nerve section, *Phil. Trans. Roy. Soc. Ser. B, 245,* 1–18

Zs.-Nagy, I., 1971, The lipochrome pigment of molluscan neurons as a specific electron acceptor, *Comp. Biochem. Physiol., 40A,* 595–602.

APPENDIX

Intelligence in a Snail[3]

By W. H. Dall

Some time since a relative told me a remarkable story about a child who had pet snails which recognized her voice and distinguished it from that of others. As such a development

[3]From the *American Naturalist,* December, 1881.

of intelligence has not hitherto been reported among mollusks, I was much interested. By the kindness of the lady from whom the story was first heard, and the intervention of Mrs. Lay, wife of Bishop Lay, formerly of Arkansas and now of Maryland, one of the family, who was cognizant of the facts, was reached, and an extract from her letter appended. I may add that Mrs. Lay speaks in the highest terms of the accuracy and clearheadedness of her correspondent (then and now a resident of Arkansas), and remarks that both she and her sister were remarkable for the ease with which they established friendly and confidential relations with the birds and animals about them. The father of these ladies, whose name I suppress merely because I have not their authority to print it, was chief clerk in the State Department under the secretaryship of Daniel Webster.

The malacologist, familiar with pulmonates, will recognize in the following quotation many facts which indicate the accuracy and unusual powers of observation of the writer. It is probable that the snail was one of the group to which *H. albolabris* belongs, at all events it was a native of Arkansas and one of the larger species. It would be highly interesting if some of our lady friends would repeat the experiment with different kinds of snails, and determine by additional evidence whether they are capable—1st. Of recognizing a call or sound; and 2d. Of distinguishing it from other similar calls or sounds; which the snail in question appears to have done.

The lady, after stating that her sister Georgie was, from the age of three years, quite an invalid and remarkable for her power of putting herself en rapport with all living things, continues: "Before she could say more than a few words, she had formed an acquaintance with a toad, which used to come from behind the log where it lived, and sit winking before her in answer to her call, and waddle back when she grew tired and told it to go away. When she was between five and six years of age, I found a snail shell, as I thought, which I gave her to amuse her, on my return from a picnic. The snail soon crawled out, to her delight, and after night disappeared, causing great lamentation. A large, old-fashioned sofa in the front hall was moved in a day or two, and on it we found the snail glued fast; it had crawled down stairs. I took a plant jar of violets and placing the snail in it carried it to her, and sunk a small top cup toy with the soil, filling it with meal. This was because I had read that French people feed snails on meal. The creature soon found it, and we observed it with interest for awhile, as we found it had a mouth which looked pink inside and appeared to us to have tiny teeth also. We grew tired of it, but Georgie's interest never flagged, and she surprised me one day by telling us that her snail knew her and would come to her when she talked to it, but would withdraw into its shell if any one else spoke. This was really so, as I saw her prove to one and another, time after time. At one time she found a number of eggs. To the best of my recollection they resembled mistletoe berries, though much whiter. They hatched, and she had fifteen or twenty little snails which used to assemble round the cup of meal which had to be frequently replenished. The old snail once fell down on to a brick pavement and its shell was fractured and a small piece lost, but Georgie pasted a piece of calico over the hole and it seemed to do very well. What became of the happy family I do not remember, nor can I tell how long my sister had them. I do not know of any more easily kept pet. If there is anything else which I have forgotten I shall be happy to write further particulars if I can recollect them."

"Georgie," my correspondent adds, "died about fourteen years ago."

An observer, who noticed and remembered the pink buccal mass, the lingual teeth and the translucent mistletoe-berry-like eggs, and after such an interval of time could so accurately describe them, is entitled to the fullest credence in other details of the story, and I have no doubt of its substantial accuracy, in spite of its surprising nature.

INDEX